Markus J. Rösch

Development of lumped element kinetic inductance detectors for mm-wave astronomy at the IRAM 30 m telescope

Karlsruher Schriftenreihe zur Supraleitung Band 012

HERAUSGEBER

Prof. Dr.-Ing. M. Noe
Prof. Dr. rer. nat. M. Siegel

Eine Übersicht über alle bisher in dieser Schriftenreihe erschienene Bände finden Sie am Ende des Buchs.

Development of lumped element kinetic inductance detectors for mm-wave astronomy at the IRAM 30 m telescope

by
Markus J. Rösch

Dissertation, Karlsruher Institut für Technologie (KIT)
Fakultät für Elektrotechnik und Informationstechnik, 2013
Hauptreferent: Prof. Dr. rer. nat. M. Siegel
Korreferenten: PD. Dr.-Ing. T. Scherer, Prof. Dr. rer. nat. A. Krabbe

Impressum

Karlsruher Institut für Technologie (KIT)
KIT Scientific Publishing
Straße am Forum 2
D-76131 Karlsruhe

KIT Scientific Publishing is a registered trademark of Karlsruhe Institute of Technology. Reprint using the book cover is not allowed.

www.ksp.kit.edu

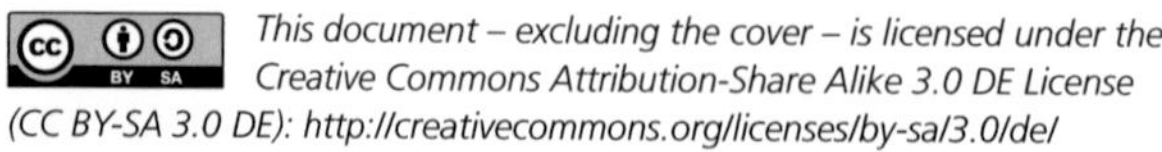

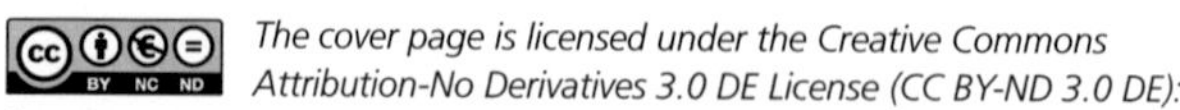

Print on Demand 2013

ISSN 1869-1765
ISBN 978-3-7315-0110-7

Development of lumped element kinetic inductance detectors for mm-wave astronomy at the IRAM 30 m telescope

Zur Erlangung des akademischen Grades eines

DOKTOR–INGENIEURS

von der Fakultät für
Elektrotechnik und Informationstechnik
des Karlsruher Instituts für Technologie (KIT)

genehmigte

DISSERTATION

von

Dipl.–Ing. Markus J. Rösch
geb. in Achern

Tag der mündlichen Prüfung:	13.06.2013
Hauptreferent:	Prof. Dr. rer. nat. M. Siegel
Korreferent 1:	PD. Dr.-Ing. T. Scherer
Korreferent 2:	Prof. Dr. rer. nat. A. Krabbe

Acknowledgment

Ich bedanke mich an dieser Stelle herzlich bei Herrn Professor Dr. rer. nat. Michael Siegel für die freundliche Übernahme des Hauptreferates. Ebenfalls bedanke ich mich bei Herrn PD. Dr.-Ing. Theo Scherer und Herrn Professor Dr. rer. nat. Alfred Krabbe für die Übernahme des Korreferates.

Mein ganz besonderer Dank gilt Herrn Dr. Karl-Friedrich Schuster, der diese interessante und herausfordende Arbeit am Institut de Radioastronomie millimétrique (IRAM) erst möglich gemacht hat und der mir immer mit fachlicher Unterstützung und wertvollen Diskussionen zur Seite stand.

I also want to thank Dr. Alessandro Monfardini, who always had an open ear for discussions and new ideas during the last 4 years. Grazie Mille!

I want to thank all my other colleagues from the Institute Néel in Grenoble for the excellent NIKA collaboration: Aurelien Bideaud, Martino Calvo, Nicolas Boudou, Alain Benoit, Francois-Xavier Désert, Christian Hoffmann and Philippe Camus.

Un grand Merci aussi aux collègues de l'IRAM pour votre aide et soutien: Roberto Garcia-Garcia, Samuel Leclercq, Arnaud Barbier, Sylvie Hallegue, Catherine Bouchet, Dominique Billon, Patrice Serres, Anne-Laure Fontana, Olivier Garnier, Francois Mattiocco et bien d'autres.

Nicht zuletzt möchte ich meinen Eltern Irmgard und Franz sowie meinen Geschwistern Christine und Katrin für die Unterstützung und Motivation während dieser Arbeit danken.

Le plus grand Merci s'adresse à la personne sans laquelle ce travail aurait été beaucoup plus difficile: ma femme Erika. Merci pour ton soutien, ta tolérance et ta compréhension pour ce que je fais.

Abstract

Since 2003, kinetic inductance detectors are considered as promising alternative to classical bolometers for mm and sub-mm astronomy. In such detectors photons with energy higher than the gap energy ($E = h\nu > 2\Delta$) of an absorbing superconductor break Cooper pairs and change the absorber's macroscopic electrical properties. The increased density of normal conducting electrons, so-called quasi-particles, changes the surface reactance of the superconductor. One possibility to measure this effect and therefore to turn the absorber into a detector is to integrate the absorber into a rf superconducting resonator circuit of high quality factor Q. The change in resonator characteristics due to the enhanced Cooper pair density can then be measured by means of standard radio frequency techniques. One particular microresonator design is the so-called Lumped Element Kinetic Inductance Detector (LEKID), which combines the resonant circuit with a sensitive direct absorption area coupled to free space. This thesis studies the development of multi-pixel LEKID arrays for the use in a large field-of-view mm-wave camera for the IRAM 30m telescope (Institut de Radioastronomie millimetrique). This includes the design and fabrication of the superconducting microresonators, the modeling and optimization of the mm-wave coupling to the detector and the characterization of multi-pixel arrays at low temperatures. The results obtained throughout this thesis brought IRAM to test a prototype instrument called NIKA (Néel-IRAM KID array) at the telescope, where first astronomical results have been achieved using this type of detector, which are also presented in this work.

Zusammenfassung

Seit 2003 werden Kinetic Inductance Detektoren als vielversprechende Alternative zu klassischen Bolometern für die mm und sub-mm Astronomie betrachtet. In solchen Detektoren brechen Photonen mit einer Energie höher als die Energielücke eines Supraleiters Cooper Paare und ändern somit die makroskopisch elektrischen Eigenschaften des Absorbers. Die erhöhte Dichte von normalleitenden Elektronen, so genannte Quasiteilchen, ändern hierbei die Oberflächenimpedanz des Supraleiters. Eine Möglichkeit, diesen Effekt messtechnisch zu erfassen bzw. den Absorber als Detektor zu nutzen, besteht darin, den Absorber in einen supraleitenden Resonator mit hoher Güte zu integrieren. Die Änderung der charakteristischen Resonatoreigenschaften aufgrund der verringerten Cooper-Paardichte kann dabei mithilfe von Standard rf Messgeräten gemessen werden. Ein spezielles Resonatordesign stellt der sogenannte Lumped Element Kinetic Inductance Detektor (LEKID) dar, der den Resonator mit einer hochempfindlichen Direkt-Absorberstruktur kombiniert, die direkt an den Freiraumwellenwiderstand gekoppelt ist. Diese Arbeit befasst sich mit der Entwicklung von Multipixel LEKID Arrays für eine mm-Wellenkamera für das IRAM 30 m Teleskop (Institut de Radioastronomie millimetrique). Dies beinhaltet das Design und die Herstellung von supraleitenden Mikroresonatoren, die Modellierung und Optimierung der mm-Wellen Einkopplung zum Detektor und die Charakterisierung von Multipixel Arrays bei kryogenen Temperaturen. Die erhaltenen Resultate dieser Arbeit erlaubten es, eine Prototypenkamera, genannt NIKA (Neel IRAM KID array), am Teleskop zu testen, wobei erste astronomische Ergebnisse mit den hier entwickelten Detektoren erzielt worden sind.

Inhaltsverzeichnis

1. Introduction

The millimeter and sub-millimeter electromagnetic spectrum is one of the frequency bands that is not efficiently used yet, but since the last decades it is subject of detailed studies. This frequency range from around 30 GHz up to several THz corresponds to emitted radiation from relatively cold objects, which makes it interesting for different applications such as medical imaging for cancer detection [2], full body scanners for security systems [38], and material science and especially for radioastronomy. The astronomy community is progressing in developing more and more powerful instruments to investigate the origins of our universe, in particular the cosmic microwave background (CMB) with a temperature of $T \approx 2.74$ Kelvin, which corresponds to emitted radiation in the mm-wave range, giving important information about the formation of the universe back to the Big Bang, 15 billion years ago. Such continuum observation instruments demand a high number of pixels and highly sensitive, usually superconducting detectors, to achieve higher resolution and mapping speed. These detectors work in general at very low temperatures, in the sub-Kelvin range, to reach high sensitivities that astronomers need for their new discoveries.

Typical low temperature mm-wave detectors are highly sensitive thermometers, so-called bolometers, measuring a temperature change caused by absorbed energy [51]. Besides semiconductor based bolometers such as MAMBO [47], one particular detector type that is being developed during the last years, is the superconducting *Transition Edge Sensor* (*TES*) [29, 32]. The bolometer detection principle of the TES takes advantage of the extremely steep slope change of resistivity vs. temperature ($\Delta R/\Delta T$) of the normal to superconducting transition (at $T \approx T_C$) and is therefore highly sensitive to small temperature changes. The TES is one of today's most sensitive bolometer based detector type for ground based continuum radioastronomy [58]. The possibility of time domain multiplexing allows the fabrication of large pixel arrays [67], which demands a readout circuit for each individual detector and therefore individual cabling as well. Working temperatures of TES detectors

are usually in the sub-Kelvin range and thus good thermal isolation of the detectors is necessary in order to reduce the power dissipation of the cryostat's cold stage. An increasing number of cables with detector count is therefore one limiting factor for large arrays. In addition, the integration of these readout circuits into the detector array is complicated and cost intensive.
A relatively new promising superconducting type of detector is the *Kinetic Inductance Detector* (*KID*) first proposed in 2003 by Caltech and JPL [15, 53, 34]. This detector does not measure the temperature change due to the absorbed energy of a photon, but measures directly the change in quasi-particle density using thin film superconducting microresonators with resonance frequencies in the range of 1-5 GHz. Theoretical calculations indicate sensitivities that are comparable to TES' but have not been demonstrated so far.
The superconducting charge carriers, called Cooper pairs, break up into normal conductive quasi-particles for absorbed energies (photons) that are higher than the superconductors energy gap ($E=h\nu>2\Delta=3.526\ k_B T_C$)(BCS-approximation) [60] leading to a change in surface impedance. The nonzero ac impedance of the superconductor due to the inertia of the Cooper pairs gives rise to an almost purely inductive reactance in addition to the conventional reactance caused by the magnetic inductance L_m, which is the so-called *kinetic inductance* L_{kin} that KIDs are famous for. The value of this L_{kin} depends on the ratio of quasi-particles to Cooper pairs and therefore on temperature, making this detector highly sensitive to small temperature changes or absorbed photons. These photons cause a change in L_{kin} and thus in surface impedance, leading to a shift in resonance frequency of the microresonator measurable by standard radio frequency techniques in amplitude and phase.
One major advantage of KIDs is the possibility to couple several hundreds of absorbing microresonators to a single transmission line to read them out by *frequency division multiplexing* (*FDM*) [39]. Hereby, a comb of frequencies corresponding to the resonance frequencies of the resonators is generated at the input. The transmitted signal is amplified by a single low noise cryogenic amplifier and is further on demultiplexed at room temperature. Compared to the TES', only one input and output transmission line (2 coax cables) running into the cryostat are sufficient to read out hundreds or thousands of detectors. This reduces the power dissipation of the cryostat's cold stage, which allows the design of larger pixel arrays. A second advantage is the relatively easy fabrication process of KIDs. The simple planar de-

tector geometry merely requires a single superconducting thin layer on, in general, a silicon substrate, which makes the fabrication of even large detector arrays easy because no junctions, bilayers or membranes are necessary. The film deposition process on the other hand has to be well understood due to a direct dependence of the detector sensitivity on homogeneity and superconducting properties of the film such as critical temperature and surface impedance.

The context of this work is the development of a KID based large field-of-view camera for the IRAM 30 m telescope in Spain, observing at 146 and 240 GHz, by the *Institute Néel* (Grenoble, France) in collaboration with the *Institut de Radioastronomie millimétrique* (*IRAM*). In the framework of this collaboration, the aim of this thesis is the design, fabrication, characterization and optimization of KID arrays and to demonstrate the evolution from the detector concept to the astronomical instrument in form of the NIKA mm-wave camera. The presented KIDs are based on a particular superconducting microresonator design, the so-called *Lumped Element Kinetic Inductance Detector* (*LEKID*) [16]. Its design allows the combination of a microwave resonator and a direct photon absorption geometry contrary to other recently developed antenna coupled KIDs [39, 5].

After a short overview about the scientific motivation behind this development and the basic concept of kinetic inductance detectors (Chapter 2), the physics, to understand these pair breaking detectors, are discussed in Chapter 3. A general resonator theory with focus on the LEKID resonant circuit is presented in Chapter 4, followed by a detailed description of the LEKID thin film deposition and microfabrication process (Chapter 5). Chapter 6 deals with the design parameters of large LEKID arrays such as resonance frequency tuning, resonator coupling and the minimization of pixel cross coupling. In Chapter 7, the modeling and optimization by calculations and simulations of the mm-wave coupling at 150 GHz is discussed and a room temperature quasi-optical measurement setup is proposed to determine the optical efficiency. Low temperature measurements of 144 pixel LEKID arrays, to determine the detector sensitivity, and first astronomical results, achieved at the IRAM 30 m telescope, are presented in Chapter 8 and 9. In order to increase the optical coupling of the detectors, a dual polarization LEKID geometry has successfully been developed and tested (see Chapter 10).

2. Scientific motivation and basic concept

2.1. The IRAM 30 meter radio-telescope

The 30 meter telescope located on Pico Veleta in the Spanish Sierra Nevada is one of the two radio astronomy facilities operated by IRAM (see Fig 2.1). It was built in 1984 at an elevation of 2850 meters and it is one of today's largest and most sensitive radio telescopes for tracing millimeter waves. The telescope is a classic single dish parabolic antenna, which allows the observation of a wide variety of objects in terms of shape, distance and nature: From solar system objects like comets to the most distant galaxies passing by star forming regions, nebula, nearby galaxies and interstellar clouds. Due to its large surface, it is unrivaled in sensitivity and is well adapted to detect faint sources. The telescope is equipped with a series of

Fig. 2.1.: The IRAM 30 m telescope at Pico Veleta in Spain

single pixel receivers operating at 3, 2, 1 and 0.8 millimeters as well as two cameras working at 1 millimeter: HERA [55] with 9 pixels for the mapping of molecular gas in extended nebulae and MAMBO, a continuum camera with 117 pixels, built by the Max-Planck-Institute for Radioastronomy (in Bonn, Germany), dedicated to the observation of dust emission from nearby molecular clouds and also out to the farthest known galaxies and black holes. The main limitations for ground based observations are due to atmospheric filtering of electromagnetic radiation. The atmosphere filters a large band of electromagnetic waves except for the visible, some near infrared frequency bands and radio-frequencies. Fig. 2.2 shows the available millimetric frequency bands at which observations can be done without suffering from a too high loss of the signal due to the atmosphere at Pico Veleta. The opacity is shown as a function of the precipitable water vapor content and the temperature of the atmosphere. We see that we have 4 frequency bands available at the Pico Veleta site, whose centers correspond to wavelengths of 0.87, 1.25, 2.05 and at 3.2 mm.

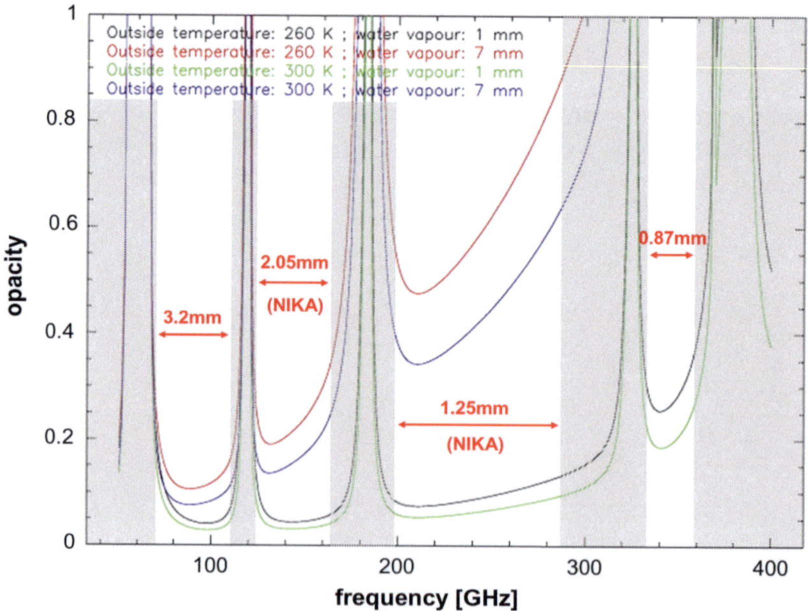

Fig. 2.2.: Available frequency bands at the IRAM 30 m telescope for observations at radio frequencies up to 400 GHz. The transparency of the atmosphere is shown dependent on the water vapor and the temperature. For each band the wavelength of the center frequency is shown.

For higher frequencies, the surface precision of the primary mirror is not sufficient anymore. The opacity is slightly different in summer and winter due to a different outside temperature, which can be seen by comparing the curves at 260 and 300 Kelvin. The most important factor is either humidity or content of precipitable water vapor in the atmosphere. Fig. 2.2 depicts the opacity for a water vapor of 1 and 7 mm.
The main interests of IRAM to develop a new camera for the two particular frequency bands at 1.25 and 2.05 mm is the investigation of high redshift objects, the cosmic microwave background, an intense study of galaxies and in particular the Sunyaev-Zeldovich effect:

2.1.1. Red shift sources

The frequency of emitted photons depends on the generation mechanism and the chemical composition of the source. The spectrum of photons, emitted due to black body radiation is described by *Planck's law* and is given by

$$B(\nu,T) = \frac{2h\nu^3}{c^2}\frac{1}{e^{\frac{h\nu}{k_B T}}-1} \quad [Wm^{-2}Hz^{-1}sr^{-1}] \tag{2.1}$$

where T is the temperature of the source, h the Planck constant, c the speed of light, k_B the Boltzmann constant, and ν the photon frequency. We know, that the universe is constantly expanding since the creation 15 billion years ago meaning that photons emitted at that time can today be observed at much lower frequencies and temperatures. An emitted photon at a time t_0 with a wavelength of λ_0, is today, at a time t, observable with a dilated wavelength due to the expansion of the universe ($\lambda(t) > \lambda_0$). The further the source is distant, relative to the observer, the older the photon and the more the wavelength is dilated. Regarding the visible spectrum, this shift in frequency corresponds to a shift to the red and is called *redshift*, where objects with this property are called *redshift sources*. The redshift is in general given by

$$z = \frac{\lambda(t)-\lambda_0}{\lambda_0} = \frac{a(t)-a(t_0)}{a(t_0)} \tag{2.2}$$

where $a(t)$ is the scaling or dilatation factor at the time t. For the opposite case, where the source is approaching the observer, this leads to an increase in frequency and we speak of blue shift. These two cases are demonstrated in Fig. 2.3a). Photons in the UV-spectrum with a temperature of around 3000 K emitted at t_0, with a redshift of around 1000 can today be observed at much lower temperatures ($\approx$3 K) and at much lower frequencies (mm wavelength range). An example of this is shown in Fig. 2.3b). A comparable effect in analogy to the expanding universe causing a dilatation in wavelength by observing moving sources is given by the famous *Doppler effect*. The fact that we can observe very distant sources at much lower frequencies and temperatures makes mm-wave astronomy that important in order to investigate such redshift sources, which would be more difficult if they were closer and visible in a much higher frequency spectrum.

2.1.2. Cosmic microwave background radiation

Among many other things, one of the most interesting mm-wave targets is the so-called *Cosmic Microwave Background* (*CMB*) radiation, first observed by Penzias and Wilson in 1965. This radiation was emitted in the earliest ages of our universe and gives important information about its creation back to the Big Bang. This CMB was emitted in a extremely hot environment of around 3000 K. Due to the expansion of the universe, we can observe this radiation today at around 2.725 K corresponding to a redshift of around 1100. Today we can observe the maximum intensity at $\lambda = 1.87$ mm (160 GHz) or in other words in the mm-wavelength range.

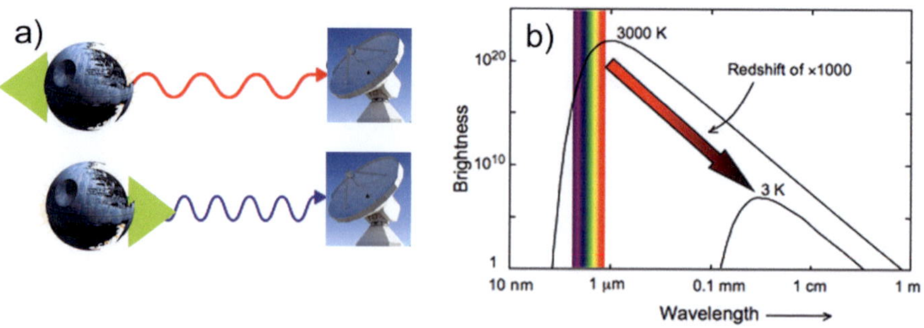

Fig. 2.3.: Principle of red and blue shifted sources (a) and the influence on the radiation spectrum and the temperature of a source (b).

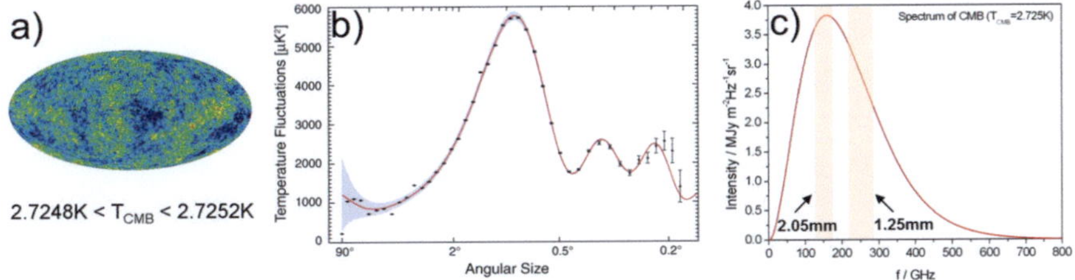

Fig. 2.4.: a) observation of the *Cosmic Microwave Background* (*CMB*) using WMAP-3years; b) temperature anisotropies in the CMB; c) spectrum of the CMB. The maximum intensity is at 160 GHz, which lies in the 2.05 mm atmospheric band.

Fig. 2.4c) shows the spectrum of the CMB with a maximum intensity in the range of the 1.25 and 2.05 mm band, making it therefore interesting and important for astronomy to observe at these particular wavelengths. Due to anisotropies the temperature of the CMB fluctuates between 2.7248 and 2.7252 K. Fig. 2.4a) shows an observation done with WMAP[1] of the CMB illustrating the temperature variations and Fig. 2.4b) the temperature fluctuations depending on the angular size. Since the discovery of the CMB, several instruments have been build to investigate it, such as COBE[2], BOOMERANG[3], ARCHEOPS [20] and WMAP. The most powerful instrument today is the PLANCK [7] satellite launched in 2009.

2.1.3. Galaxies and Sunyaev-Zel'dovich effect

Another temperature fluctuation effect observable at mm-waves is caused by inverse Compton scattering of high energy electrons distorting the CMB radiation and is described by the famous $Sunyaev-Zel'dovich$ ($SZ-$) effect. Low energy photons of the CMB are energized by colliding with these high energy electrons often found in hot gases within galaxy clusters and receive therefore additional energy. Direct observations of such clusters is difficult, because the hot gases and dust clouds filter much of the radiation we are interested in to study. But it is instead possible to investigate such sources using the SZ-effect by observing the distortions of the CMB around them. To observe these temperature fluctuations in large galaxy clusters, a

[1] Wilkinson Microwave Anisotropy Probe [6]

[2] Cosmic Background Explorer [9]

[3] Balloon Obervations of Millimetric Extragalactic Radiation and Geophysics [21]

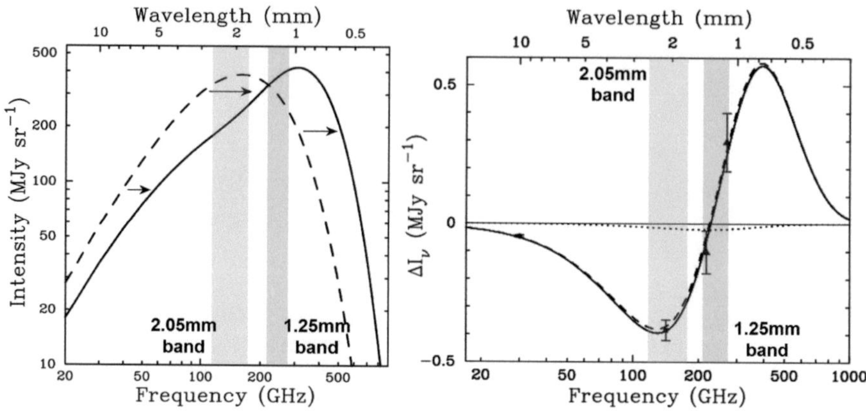

Fig. 2.5.: [12] Principle of the distortion of the cosmic microwave background due to the Sunyaev-Zel'dovich effect (left) and the intensity fluctuations at different wavelengths (right). The yellow and orange bars indicate the 1.25 and 2.05 mm band.

higher angular scale is necessary compared to the observations of the CMB anisotropies. Taking the Planck satellite as an example, its angular scale is too small because of the size of its primary mirror (1.5 m). Ground based telescopes such as the IRAM 30 m with large angular scale are more suitable for this kind of observations and therefore among other reasons interesting for mm-wave astronomy. Fig. 2.5 demonstrates why these observations are particularly important in the mm-wave range. On the left hand side, the intensity of the CMB (dashed line) is compared to the distorted intensity (solid line) due to the SZ-effect and on the right hand side at which frequencies we can observe the highest fluctuations. Not only is the maximum intensity measurable at mm-wavelengths, in particular at 1 and 2 mm, but also the highest intensity fluctuations can be found in this range.

2.2. NIKA: A dual band kinetic inductance camera for the IRAM 30 m telescope

NIKA stands for the *Néel IRAM KID ARRAY*, a collaboration that was formed between several institutes for the development of a large field-of-view camera for the IRAM 30 m telescope based on kinetic inductance detectors (KIDs). The camera works simultaneously at 1.25 and 2.05 mm corresponding to the frequencies of 240 and 146 GHz respectively (see above). The 2.05 mm array is being developed by the institute *Néel* and IRAM in Grenoble (France) in collaboration with the Cardiff University (UK). The detectors for the 2.05 mm array are so-called *Lumped Element Kinetic Inductance Detectors* (*LEKID*), whose development and optimization for the NIKA instrument is discussed in detail throughout this thesis. The detector array for the 1.25 mm band is being developed at the *Space Research Institute of the Netherlands* (SRON) based on antenna coupled MKIDs [5] and will not be described here.

After the formation of the NIKA collaboration at the end of 2008, the development of LEKIDs for large field array detectors started on the basis of the previous work done by Doyle during his dissertation [16]. Doyle describes in detail the theory of such lumped element kinetic inductance detectors for the far infrared, their use as direct absorber and shows low temperature measurements of single aluminum resonators. Since then, starting with six pixel arrays to investigate resonator properties, the development went fairly rapidly toward a KID based multi-pixel camera for the IRAM 30 m telescope. Within only three years, the successful development of detector arrays, the camera and the readout electronics allowed to test three prototype instruments at the telescope, each with significantly improved performances. In 2009, NIKA demonstrated for the first time the imaging capabilities of LEKIDs. We have improved the sensitivity and the optical efficiency of the LEKIDs by important factors compared to the first results and are today competitive with existing bolometer based instruments such as MAMBO and GISMO. NIKA is today the most sensitive KID based instrument at the 1.25 and the 2.05 mm band.

The future NIKA instrument is foreseen to be installed at the telescope in 2014, allowing observations at the two mentioned frequency bands simultaneously with a field of view of 6 arcmin. The pixel count is 1000 for the 2.05 mm band and 3000 for the 1.25 mm band. The aimed sensitivity, given in *noise equivalent power* (NEP) and *noise equivalent temperature* (NET), is around $NEP = 6 \cdot 10^{-17}\ W/Hz^{0.5}$ and $NET = 0.7\ mK/Hz^{0.5}$ at 2.05 mm and $NEP = 6 \cdot 10^{-17}\ W/Hz^{0.5}$ and $NET = 0.8\ mK/Hz^{0.5}$ at 1.25 mm. This instrument, in its final configuration, will be a state of the art camera for mm-wave astronomy.

2.3. Working principle of kinetic inductance detectors

Since 2003 kinetic inductance detectors are considered as promising alternative to classical bolometers for mm and sub-mm astronomy [15]. In such detectors photons with energy higher than the gap energy ($E = h\nu > 2\Delta$) of an absorbing superconductor break cooper pairs and change the absorber's macroscopic electrical properties. The increased density of quasi-particles changes the surface reactance of the superconductor (kinetic inductance effect [60]). Quasi-particles are normal conducting electrons excited due to incoming photons. One possibility to measure this effect and therefore to turn the absorber into a detector is to integrate the absorber into a rf superconducting resonator circuit of high quality. The change in resonator characteristics due to the reduced cooper pair density can then be measured by means of standard radio frequency techniques. Usually the resonator is loosely coupled to a transmission line and the measurement of the complex transmission scattering parameter S_{21} is performed. The principle of coupling resonators electromagnetically to a transmission line is shown in Fig 2.6 for the case of an inductive coupling. An illumination of the excited detector leads to a shift in resonance frequency (f_0), in amplitude (A) and in phase (ϕ). Fig. 2.7 shows the S_{21} of an array of 25 resonators coupled to one transmission line (left hand side). On the right hand side the detection principle of KIDs is illustrated. It shows a zoom on one of the 25 resonators for

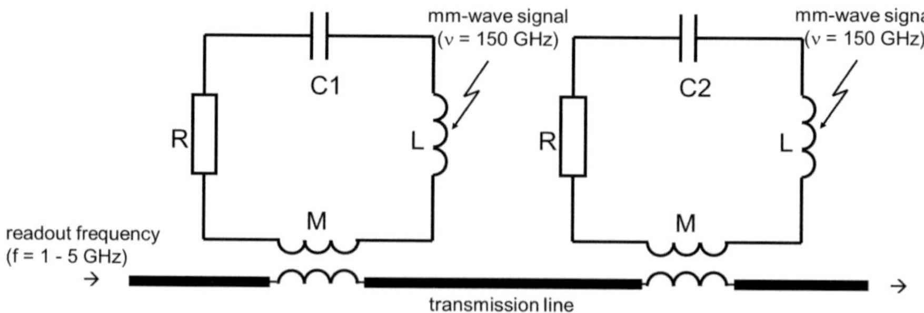

Fig. 2.6.: Two resonators with resonance frequency f_1 and f_2 coupled inductively over a mutual inductance M to a transmission line. The inductive part of the resonator is unchanged in this case and the resonance frequencies are determined by the capacitors C_1 and C_2. The resonance frequencies of the resonators is typically in the range of $f = 1$ - $5\ GHz$, whereas the frequency ν of the incoming photons is at much higher frequencies, in our case at $\nu \approx 150\ GHz$.

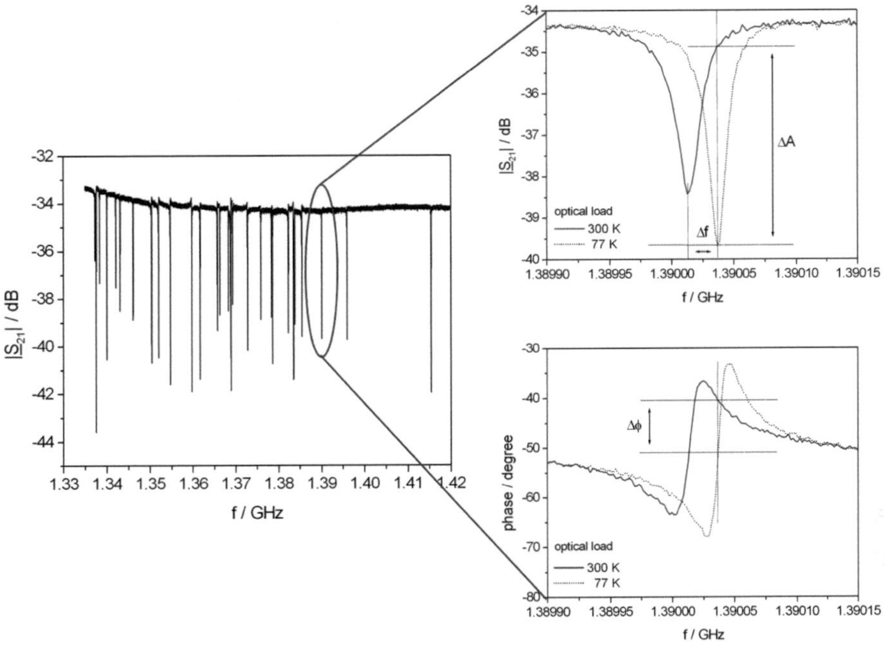

Fig. 2.7.: Left: Scan of the transmission scattering parameter S_{21} of a 25 pixel KID array; right top: Magnitude of S_{21} in dB over frequency; right bottom: phase in degree over frequency.

two different optical loads (77 and 300 Kelvin). Under higher optical loading (300 Kelvin), the resonance frequency shifts downwards due to the increased number of quasi-particles, which increases the impedance of the resonator leading to a change in amplitude and phase. The complex scattering parameter S_{21} at each resonance can also be visualized in a complex IQ-plane, where I is the inphase component and Q the quadrature component or real and imaginary part of S_{21}. This is shown in Fig. 2.8. In this formalism the change in amplitude and phase can directly be visualized in one single plane. It also shows, that the phase response is bigger compared to the response in amplitude, which is important for the readout of such detectors. This scheme naturally is well adapted to a frequency-domain multiplexing allowing the readout of a large number of resonators. Grouped within a limited bandwidth, a single transmission line is sufficient to read out several hundreds of pixels [69, 59].

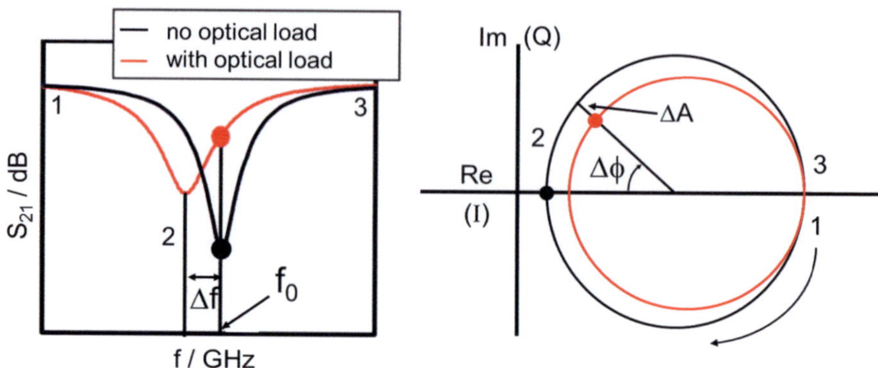

Fig. 2.8.: Demonstration of the signal detection measurable in amplitude (ΔA) and phase ($\Delta \phi$) by plotting S_{21} in a complex IQ-plane.

2.3.1. Lumped element kinetic inductance detectors

In Fig. 2.9, a particular resonator design, the so-called Lumped Element Kinetic Inductance Detector (LEKID) is shown. The implementation of this resonator type as kinetic inductance detector was first proposed by Doyle in 2007 [17]. A high and constant current density over the whole length of the meander assures that this part is sensitive to direct detection. Compared to alternative microwave resonators with distributed character e.g. antenna coupled quarter wavelength resonators [39, 5], this type consists of a long meandered line, the inductive part, and a relatively compact interdigital capacitor. In other words, the LEKID combines the resonator with an absorbing antenna in the same structure. The geometry of the meander is chosen to achieve maximum optical efficiency. The design is fully planar and merely requires a back-short cavity situated at a calculated distance behind the resonator. The relatively dense inductor meander structure creates a square shaped global pixel absorber area, which can be densely packed. Therefore, the LEKID absorber-resonators can be arranged in a filled array geometry avoiding lenses or antenna structures for individual pixels. The resonance frequency can be tuned by changing the length and the number of the fingers of the interdigital capacitor.

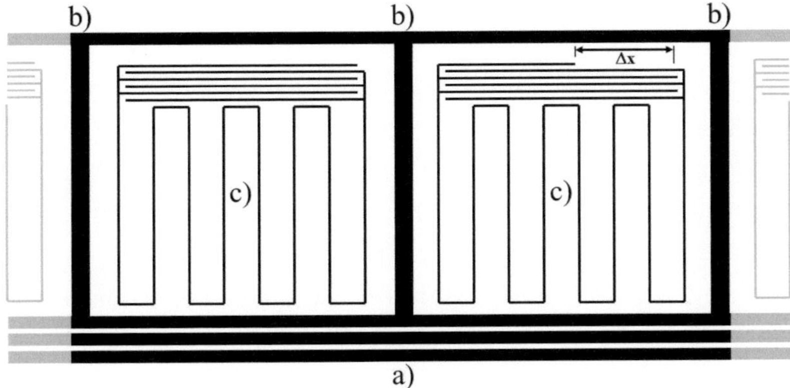

Fig. 2.9.: Two lumped element kinetic inductance detectors (LEKID) (c) inductively coupled to a coplanar waveguide transmission line (a). They consist of a meandered line, the inductive part, and an interdigital capacitor. The resonance frequency can be tuned by changing the length of the interdigital capacitor fingers by Δx. The two resonators are surrounded by a ground plane (b) to decrease cross-coupling effects between them.

2.4. Problem assignment and focus of this thesis

The development of large detector arrays with several hundreds of pixels coupled to and read out over a single transmission line represents a challenge. The advantage of multiplexing hundreds of detectors demands today's most sophisticated FPGAs and analog to digital converters that provide sufficient large bandwidth. The current NIKA readout electronics allow the packaging of around 115 resonators with 2 MHz spacing in a 230 MHz bandwidth. A frequency spacing of only 2MHz between each detector is only possible for resonators with high quality factors in the range of $Q = 100000$ and a well known frequency tuning method. To achieve such quality factors, a deposition and microfabrication process that guaranties homogeneous superconducting films with reproducible properties, needs to be controlled and optimized. The homogeneity of the film thickness is one important factor, since the kinetic inductance can change significantly with small variations in thickness, as we will see later.

The combination of resonator structure and direct absorption area is one big advantage of LEKIDs, but both are difficult to optimize independently. For a optimal mm-wave coupling of the detector, the meander geometry cannot be chosen ran-

domly as it is the same for the determination of the resonance frequency of each resonator. For mm-wave absorption at $\lambda = 2.05$ mm, the pixel size needs to be of the same order with a calculated distance between the lines, which limits the meander length and thus the line inductance to determine the resonance frequency. To define a resonance frequency without lowering the mm-wave absorption, the combination of interdigital capacitor and inductive meander line needs to be well designed as a compromise between sufficient mm-wave coupling and high quality resonators. One of the biggest problems that arises by designing multi-pixel arrays that are relatively densely packed on a limited wafer area is the electromagnetic cross coupling between resonators. This cross coupling has big influence on the resonance frequency of each resonator and can therefore lead to imaging errors by causing false detector responsivities. It depends on the geometrical distribution of each pixel on the array and the distance between resonators.
Within the framework of the NIKA collaboration, the main focus of this thesis is directed to the fabrication, the design and measurement of multi-pixel arrays and the modeling and optimization of the optical efficiency primarily for the 2.05 mm band:

Film deposition and microscale processing: The superconducting properties of the LEKID material is one of the key factors for a highly sensitive KID and depend strongly on the material quality and thus on the fabrication process. To guarantee a high and constant pixel sensitivity over an entire detector array, a homogeneous superconducting film with thickness in the range of d=20-30 nm has to be deposited on a substrate. The homogeneity and the quality of such films are directly linked to the kinetic inductance and therefore to parameters such as resonance frequency and detector responsivity. In Chapter 5, two different deposition methods are discussed and the film quality and homogeneity are compared. At high frequencies the quality of the structure edges can lead to increased losses in the resonator, which degrades the performance of the detector, making the microprocessing an important fabrication step. In the same chapter, the fabrication process, in particular the lithography and etching step, is demonstrated in order to achieve high quality resonator structures.

Designing and measuring LEKID arrays: The design of multi-pixel plane field arrays not only demands high resonator quality factors but also a high number of densely packed resonators in a limited bandwidth. The determination of the resonance frequency for each resonator with a constant spacing between them is therefore one important factor for building arrays with several hundreds of pixels. The geometrical distribution of the resonators over the array is another design parameter that has to be taken into account. To pack a large number of pixels onto a given area, the electromagnetic coupling between resonators, influencing the resonance frequency of neighboring pixels, has to be understood and minimized. Chapter 6 deals with the design and optimization of resonator properties such as the coupling to the transmission line, the frequency tuning and the reduction of electromagnetic cross coupling between pixels.

Modeling and measuring the mm-wave coupling efficiency: To achieve high detectivity for this direct absorption detectors, the optical efficiency has to be sufficiently high. This means that the detector impedance has to be matched to the free space wave impedance to obtain maximum mm-wave coupling. A possibility to model this coupling analytically and also to perform numerical simulations is demonstrated in Chapter 7 and compared to measurements done with a room temperature quasi-optical mm-wave reflection measurement setup. This setup is used to investigate and to optimize the power absorption in LEKIDs, where the results can directly be transferred to the cryogenic case.

Dual polarization pixel design: To further increase the optical efficiency, new LEKID geometries can offer a higher mm-wave coupling, where more power is absorbed compared to the classical meander structure. One of the main topics is the investigation of a LEKID geometry that is sensitive to two polarizations to increase the optical efficiency. One possible structure is presented in Chapter 10, where the pixel design is characterized electrically as well as optically by simulations and by room temperature and cryogenic measurements.

3. Electrodynamics of superconductors, the kinetic inductance and sensitivity limits

The detectors presented in this work are based on the kinetic inductance effect [60] in superconductors, which can be referred to a change in surface impedance due to an increased temperature or photons with sufficient energy hitting the superconductor. The functionality of this kind of detector depends on several superconducting properties such as the critical temperature T_C and the surface impedance $\underline{Z}_s$, both determining the responsivity of the detector. These parameters have to be well understood in order to design and fabricate a highly sensitive detector device. This chapter will therefore give an overview to the physical background of these detectors, focusing in particular on the working principle and the fundamental superconducting properties of the kinetic inductance detectors presented in this thesis.

3.1. The complex conductivity and the surface resistance

The property for which a superconductor is famous for, is the vanishing dc resistance at a temperature below the critical temperature T_C. At these temperatures the conducting electrons of a superconductor form so-called *Cooper pairs*, the charge carriers providing the supercurrent that are bound together with a binding energy of $2\Delta \approx 3.526 k_B T_C$ [60]. This, however, is different for ac currents at high frequencies. In this case, the real part of the surface impedance is not zero, but depends on the number of quasi-particles in the superconductor. *Quasi-particles* are normal conducting electrons thermally excited due to incoming photons or simple heating of the superconductor. A Cooper pair can be broken up into two quasi-particles by an energy that is two times higher than the gap energy of the superconductor ($E = h\nu > 2\Delta$). The ratio of the density of normal conducting electrons n_n and superconducting Cooper pairs n_s depends on the temperature and the frequency. The number of quasi-particles increases hereby for increasing temperature and increasing frequency changing the surface impedance accordingly. We can express this effect by introducing a complex conductivity according to the *two fluid model* (see Appendix A.1), which is given by

$$\sigma = \sigma_1(n_n) - j(\sigma_2(n_s) + \sigma_2(n_n)) \tag{3.1}$$

with $j = \sqrt{-1}$. The real and imaginary part of this conductivity can be expressed by

$$\sigma_1 = \frac{n_n e^2 \tau}{m(1+\omega^2\tau^2)} \tag{3.2}$$

and

$$\sigma_2 = \frac{n_n e^2 (\omega\tau)^2}{m_e(1+\omega^2\tau^2)} + \frac{n_s e^2}{m_e \omega} \tag{3.3}$$

where ω is the angular frequency, τ the electron-phonon relaxation time, e the charge of an electron and m_e the effective mass of an electron. For the superconducting carriers, we can assume that $\tau \rightarrow \infty$ because there is no scattering effects with the lattice, which reduces the total conductivity to

$$\sigma = \sigma_1 - j\sigma_2 = \frac{n_n e^2 \tau}{m(1+\omega^2\tau^2)} - j\frac{n_s e^2}{m\omega} \tag{3.4}$$

This conductivity depends on the material properties of the superconductor (n_s, n_n, τ), the frequency ω, and the temperature T, where the T-dependence follows from the *London penetration depth* $\lambda_L(T)$. Expressing the complex conductivity with $\lambda_L(T)$ (see Appendix A.1) gives

$$\sigma = \sigma_1 - j\sigma_2 = \frac{n_n}{n}\sigma_n - j\frac{1}{\omega\mu_0\lambda_L^2(T)} \tag{3.5}$$

where $\lambda(T=0) = \sqrt{m_e/\mu_0 n_s e^2}$. With the complex conductivity derived above we can calculate the high frequency surface impedance of a superconducting film. In general, it is given by

$$\underline{Z}_s = \frac{\int \vec{E} d\vec{s}}{\int \vec{H} d\vec{s}} = \sqrt{\frac{j\omega\mu_0}{\sigma}} \tag{3.6}$$

assuming a harmonic plane wave. Inserting 3.5 in 3.6, we get the surface impedance dependent of the complex conductivity from the two fluid model:

$$\underline{Z}_s = \sqrt{\frac{j\omega\mu_0}{\frac{n_n}{n}\sigma_n - j\frac{1}{\omega\mu_0\lambda_L(T)^2}}} = R_s + jX_s \tag{3.7}$$

which can be approximated with

$$\underline{Z}_s \approx \frac{1}{2}\omega^2\mu_0^2\lambda_L^3(T)\sigma_1 + j\omega\mu_0\lambda_L(T) \tag{3.8}$$

The imaginary part of this surface resistance directly depends on the density of cooper pairs and the temperature, both combined in the temperature dependent London penetration depth $\lambda_L(T)$. With $\sigma_2 = n_s e^2/m\omega$ and $\lambda_L(T)$ we can express X_s by

$$X_s = \frac{\sqrt{\frac{\omega\mu_0}{\sigma_2}}}{\sqrt{1-(\frac{T}{T_C})^4}} \tag{3.9}$$

This way, it depends directly on σ_2.
For the particular case of very thin superconducting films with thickness d, where $d << \lambda_L$ and the mean free path of electrons l_{mfp} is limited by d, the main losses are due to surface scattering and the current density can be considered as homogeneous in the entire film cross section [22]. This reduces $\underline{Z}_s$ to

$$\underline{Z}_s = \frac{1}{(\sigma_1 - j\sigma_2)d} \tag{3.10}$$

For the superconducting films discussed in this work, equation (3.10) is used to calculate the surface impedance. Later we will see that thinner films, that fulfill the condition $d << \lambda_L$, lead to a higher responsivity of the detector.

3.2. The kinetic Inductance

In a superconducting strip with an applied electrical field in the form of $E = E_0 \cdot e^{j\omega t}$ energy can be stored in two different ways. One is the classical magnetic energy E_{mag} due to the magnetic field generated by the applied current, which depends on the geometry of the conductor line. This magnetic energy is given by

$$E_m = \int \frac{\mu_0 \vec{H}^2}{2} dV = \frac{1}{2} L_m I^2 \tag{3.11}$$

where L_m is the magnetic Inductance consisting of

$$L_m = L_{m,int} + L_{m,ext} \tag{3.12}$$

$L_{m,int}$ represents the inductance due to the magnetic field that penetrates, dependent of λ_L, into the superconductor and $L_{m,ext}$ is the inductance that arises from the magnetic field around the conductor and is defined by the length and the cross section of the conductor.

A second part of the energy can be stored as kinetic energy E_{kin} for which effect the kinetic inductance detectors are famous for. As described by the *two fluid model*, the charge carriers of a superconductor consist of Cooper pairs and a certain amount of excited quasi-particles depending on the temperature. The Cooper pairs possess inertia because they cannot follow the ac-current frequency instantaneously leading to a large reactance at high frequencies and making the supercurrent J_s lagging the electrical field E by $90°$, which shows that the reactance due to the ac-current is purely inductive. Therefore we can write in general for the kinetic energy E_{kin} of the Cooper pairs with $J_s = n_s e v_0$:

$$E_{kin} = \int \frac{m_e n_s v_0^2}{2} dV = \frac{m_e}{2 n_s e^2} \int J_s^2 dV = \frac{1}{2} L_{kin} I^2 \tag{3.13}$$

where v_0 is the average velocity of the charge carriers and L_{kin} is the famous *kinetic inductance*.

From 3.13 we can see that for an increase in density of Cooper pairs n_s by decreasing the temperature, the kinetic energy decreases and reaches its minimum at

$T = 0K$. For increasing temperature, n_s decreases due to quasi-particle excitation until the critical temperature T_C is reached and the superconductor becomes normal conductive. At $T = T_C$ all Cooper pairs are broken up to quasi-particles and the losses are dominated by the electron-phonon scattering. We can explain this with the schematic of the complex conductivity and refer it to the kinetic energy E_{kin}. The current J consists in that case of two different parts, the superconducting part J_s and the normal conducting part J_n ($J = J_s + J_n$). At a temperature of $T = 0K$ we have only Cooper pairs and no quasi-particles as charge carriers ($J = J_s$). In that case n_s and σ_2 reach their maximum value to provide the supercurrent J_s. By increasing the temperature, Cooper pairs break up into quasi-particles leading to a decrease of n_s and σ_2 and an increase of n_n and σ_1. A lower Cooper pair density forces the Cooper pairs to increase their velocity in order to provide the same supercurrent J_s. This means, that the reactive part of the surface impedance $\underline{Z}_s$ increases due to the inertia of the Cooper pairs, giving rise to the kinetic inductance L_{kin}. The real part of the surface impedance, on the other hand, increases due to the current J_n provided by the normal conducting charge carriers (quasi-particles).
This shows, that the kinetic energy, due to the inertia of the Cooper pairs, varies with the reactance of $\underline{Z}_s$ and therefore with the kinetic inductance defined above. L_{kin} depends not only on the volume of the conductor but also on the density of Cooper pairs and therefore on the temperature. Remember that the magnetic inductance only depends on the geometry of the conductor. The effect of a small variation in temperature leading to a change in the surface impedance due to change in kinetic inductance is the basis of the detectors presented in this thesis. Depending on the material, the kinetic inductance is therefore dominating the geometrical inductance at low temperatures. The fraction of kinetic and geometrical inductance defines thus the responsivity of the detector and is called the kinetic inductance fraction $\alpha = L_{kin}/(L_{kin} + L_m)$. The higher this value, the higher the expected detector response. For typical aluminum thin films, we determine values of $\alpha \approx 0.2$.
Seen, that L_{kin} can directly be referred to the surface impedance $\underline{Z}_s$, we can write

$$\underline{Z}_s = R_s + jX_s = R_s + j\omega L_{kin} \tag{3.14}$$

In Chapter 3.1 the calculation of $\underline{Z}_s$ for superconductors was derived and we can directly calculate the kinetic inductance. The superconducting films used in this work are thin compared to the London penetration depth ($d << \lambda_L$). We can therefore use (3.10) to calculate $\underline{Z}_s$ and derive the kinetic inductance with

$$L_{kin} = \frac{\sigma_2}{d(\sigma_1^2 + \sigma_2^2)\omega} \tag{3.15}$$

The calculated kinetic inductance L_{kin} using (3.15) is given in H/square. Another way to derive the kinetic inductance is described in detail in [16].

3.3. Consequences of the Pippard non-locality and the Mattis-Bardeen theory

In 1953, Pippard introduced a non-local treatment of superconductors with $\xi_0 >> \lambda_L$ [48], where ξ_0 is the average length over which the electrons, forming a Cooper pair, can interact (see A.1). This is based on the non-local generalization of Ohm's law presented by Chambers [13]. Due to the large coherence length, compared to the London penetration depth, in many superconductors such as aluminum, the rapidly varying electrical field cannot anymore be considered as constant over the length of ξ_0. Superconductors with this properties are therefore called Pippard-superconductors. The non-locality of Pippards theory gives rise to an effective London penetration depth λ_{eff}, which can be significantly larger than $\lambda_L(T)$. It can be approximated with [60, 63] to

$$\lambda_{eff} = 0.65(\lambda_L^2(T)\xi_0)^{1/3} \tag{3.16}$$

This gives for aluminum a $\lambda_{eff} \approx 50nm$ with $\lambda_L(0) = 16nm$. For very thin films where l_{mpf} the mean free path of the electrons is limited by the film thickness d, λ_{eff} can be expressed by

$$\lambda_{eff} = \frac{4}{3}\sqrt{\frac{\xi_0}{d}} \tag{3.17}$$

The increased penetration depth λ_{eff} changes the complex conductivity introduced in (3.1). Due to the dependency of σ_2 from λ_L, the imaginary part of the surface impedance increases to much higher values leading to a higher kinetic inductance. It were Mattis and Bardeen [37] who applied the Pippard non-local theory and the temperature dependent gap energy to the surface impedance of a superconductor. They calculated the complex conductivity dependent of the temperature, the frequency, the gap energy and the normal conductivity of the conductor σ_n just above the critical temperature T_C. The integrals they derived for the complex conductivity are:

$$\begin{aligned}\frac{\sigma_1}{\sigma_n} &= \frac{2}{\hbar\omega}\int_{\Delta}^{\infty}[f(E)-f(E+\hbar\omega)]\,g(E)dE \\ &\quad +\frac{1}{\hbar\omega}\int_{\Delta-\hbar\omega}^{-\Delta}[1-2f(E+\hbar\omega)]\,g(E)dE\end{aligned} \tag{3.18}$$

$$\frac{\sigma_2}{\sigma_n} = \frac{1}{\hbar\omega}\int_{\Delta-\hbar\omega,-\Delta}^{\Delta}\frac{[1-2f(E+\hbar\omega)]\,(E^2+\Delta^2+\hbar\omega E)}{\sqrt{(\Delta^2-E^2)\left[(E+\hbar\omega)^2-\Delta^2\right]}}dE \tag{3.19}$$

where $f(\eta)$ is the Fermi function

$$f(\eta) = \frac{1}{1+e^{\frac{\eta}{k_BT}}} \tag{3.20}$$

and $g(E)$ is given by

$$g(E) = \frac{E^2+\Delta^2+\hbar\omega E}{\sqrt{(E^2-\Delta^2)\left[(E+\hbar\omega)^2-\Delta^2\right]}} \tag{3.21}$$

The lower limits of the σ_2-integral depend on the frequency range and have to be used as follows: for $\hbar\omega << \Delta \rightarrow \Delta-\hbar\omega$ and for $\hbar\omega >> \Delta \rightarrow -\Delta$. The second integral of σ_1 vanishes for $\hbar\omega << \Delta$ and $k_BT << \Delta$ and the integrals can be reduced to [24]

$$\frac{\sigma_1}{\sigma_n} = \frac{2\Delta(T)}{\hbar\omega}e^{\frac{\Delta(0)}{k_BT}}K_0\left(\frac{\hbar\omega}{2k_BT}\right)\left[2sinh\left(\frac{\hbar\omega}{2k_BT}\right)\right] \tag{3.22}$$

$$\frac{\sigma_2}{\sigma_n} = \frac{\pi\Delta(T)}{\hbar\omega}\left[1 - 2e^{\frac{-\Delta(0)}{k_B T}}\, e^{\frac{-\hbar\omega}{2k_B T}} I_0\left(\frac{\hbar\omega}{2k_B T}\right)\right] \tag{3.23}$$

where K_0 and I_0 are modified Bessel-functions of the first and second kind. With this complex conductivity for Pippard superconductors such as aluminum, the surface impedance $\underline{Z}_s$ can be calculated as described in (3.1). For $T << T_C$, $\sigma_1 << \sigma_2$ and therefore $R_s << \omega L_s$ [16] leading to a detector device that is highly sensitive to small changes in kinetic inductance caused by heating the detector or by absorbing photons with $E > 2\Delta$. This effect is therefore perfectly adapted by using a superconducting resonant circuit as detector, where the change in kinetic inductance leads to a variation in resonance frequency of the resonator, which can relatively easily be measured.

3.4. Thermal and photon quasi-particle excitation

For temperatures $T << T_C$, the number of thermally excited quasi-particles can be approximated by [39]

$$N_{qp} = 2N(0)vol\sqrt{2\pi k_B T\Delta(0)}\; e^{\frac{-\Delta(0)}{k_B T}} \tag{3.24}$$

where $N(0)$ is the single spin density of electron states at the Fermi surface and vol is the volume of the superconductor (see appendix A.1). This equation does not take into account the quasi-particles that are excited due to photons with $E > 2\Delta$. If a superconductor absorbs such photons the number of quasi-particles increases until it reaches an equilibrium value where the created number of quasi-particles per unit time is equal to the number of recombining electrons to Cooper pairs. This time is called the quasi-particle life time τ_{qp} and it depends on the film's thickness and the purity of the material. A theoretical value for τ_{qp} is given by [30]

$$\tau_{qp} = \frac{\tau_0}{\sqrt{\pi}}\left(\frac{k_B T_C}{2\Delta}\right)^{\frac{5}{2}}\left(\frac{T_C}{T}\right)^{\frac{1}{2}} e^{\frac{\Delta}{k_B T}} \tag{3.25}$$

where τ_0 is a material dependent parameter in the order of $438 \cdot 10^{-9}s$ [30] for aluminum. We can see that the quasi-particle lifetime also depends on temperature and reaches its maximum for $T \to 0K$. Combining now (3.24) and (3.25), we can express τ_{qp} depending on the quasi-particle density n_n in the superconductor.

$$\tau_{qp} = \frac{\tau_0}{n_n} \frac{N(0)(k_B T_C)^3}{2\Delta^2} \tag{3.26}$$

Here we see that τ_{qp} is inversely proportional to the quasi-particle density. To achieve therefore relatively long lifetimes, the superconducting film has to be cooled to temperatures far below T_C. The interest in having long lifetimes is due to the excess of quasi-particles due to a continuous flux of photons with $E > 2\Delta$, which depends also on τ_{qp}. The number of these quasi-particles $N_{qp(photon)}$ can be estimated by

$$N_{qp(photon)} = \frac{\eta P_{photon} \tau_{qp}}{\Delta} \tag{3.27}$$

where P_{photon} is the power of the photon flux and η is a material dependent efficiency factor. This proves that the number of excited quasi-particles is proportional to the lifetime τ_{qp} giving the possibility by lowering the temperature to $T << T_C$ to increase the responsivity of the KID.

3.5. Fundamental noise limits of KIDs

Generation recombination noise A frequently used figure to describe the sensitivity of a detector is the *noise equivalent power* (*NEP*) which is defined as the input power of a system required to obtain a signal to noise ratio (SNR) equal to 1 in a 1Hz bandwidth. The fundamental noise limit of KIDs is determined by the generation recombination noise caused by statistical fluctuations in the quasi-particle and Cooper pair density for $T > 0$. This can be expressed in NEP and can be calculated with [56]

$$NEP_{g-r} = 2\Delta\sqrt{\frac{N_{qp}}{\tau_{qp}}} \propto e^{\frac{-\Delta(0)}{k_B T}} \quad [W/Hz^{0.5}] \tag{3.28}$$

As we can see, the NEP is proportional to the square root of the number of quasi-particles in a given film volume. To reduce N_{qp} and therefore the generation recombination noise, we can reduce either the temperature or the film volume. As mentioned before, a lower film volume also increases the kinetic inductance of the superconductor to achieve higher responsivity. Thus it is important for two reasons, both the kinetic inductance and the noise, to use very thin superconducting films for a highly sensitive KID. This value is usually in the order of $NEP \approx 10^{-18} W/\sqrt{Hz}$ for such kinetic inductance detectors. This number does not take into account the noise that is added to the detector due to absorbed photons and is therefore the lower limit of noise.

The photon noise The second fundamental noise limit is the so-called photon noise or photon shot noise depending on the number and bandwidth of the incoming background photons absorbed in the detector. Observing an astronomical source by looking at the sky always means that there is also a certain power coming from the atmospheric background that is detected, however, it is no part of the actual signal. This power (from source and background) adds noise to the detector and is limiting the optical sensitivity. The photon noise is also expressed in NEP and is given by

$$NEP^2 = NEP^2_{photon} + NEP^2_{phonon} = 2h\nu P_{pixel} + 2k_B T_{equ} P_{pixel} \tag{3.29}$$

where T_{equ} is the equivalent temperature that the detector sees dependent on the atmospheric and the telescope temperature and the telescope transmission coefficient. P_{pixel} is the power hitting one pixel of the detector array. This NEP is separated in two parts, one the photon noise NEP_{photon} and the second one the phonon noise NEP_{phonon}. The second part arises due to the fact that photons behave as bosons that underlie the Bose-Einstein law adding a contribution to the total noise. In the case of the IRAM 30 m telescope, we can estimate a value of $T_{equ} \approx 16.5\ K$. Calculating the corresponding NEP, we see that the NEP_{phonon} is roughly of the same order than NEP_{photon}. This is not the same for all frequency ranges but for the NIKA 1.25 and 2.05 mm band it can be assumed to be in the same order. We can therefore write as a very rough estimation for the total photon noise the following expression:

$$NEP_{photon} \approx \sqrt{4h\nu P_{pixel}} \quad [W/Hz^{0.5}] \tag{3.30}$$

For a typical background power of $P_{pixel} = 9\ pW$ hitting the detector at ν=150 GHz in the laboratory or at the telescope, we calculate $NEP_{photon} \approx 6 \cdot 10^{-17}\ W/Hz^{0.5}$. This is higher than the formerly introduced generation recombination noise, which means that the sensitivity of the LEKIDs under optical loading is limited by the photon noise. This photon noise, linked to the power of the astronomical source, is therefore the physical sensitivity limit of the NIKA instrument which cannot be exceeded. To achieve this sensitivity limit, all other potential noise sources have to be decreased below the photon noise. This includes the noise caused by the instrument itself such as the electronics and the detector noise, the cryogenic amplifier noise and noise coming from the environment such as parasitic magnetic field effects.

4. Lumped Element Kinetic Inductance Detectors

As discussed in the previous chapters, photons with $h\nu > 2\Delta$ break Cooper pairs into quasi-particles in a superconducting film leading to a change in the surface impedance of the superconductor. The kinetic inductance detector consists of a superconducting resonant circuit, a microresonator, where the change in surface impedance due to incident photons, affects the properties of the resonator. A higher or lower surface impedance can directly be measured by the change in amplitude, phase and resonance frequency. The performance of the KID depends strongly on the design of these resonant circuits. In this chapter we present the design and the possibilities to model one of these resonant circuits, the so-called *Lumped Element Kinetic Inductance Detector* (*LEKID*), which was first proposed by Doyle [16] as a feasible alternative to the distributed antenna coupled quarter wavelength resonators.

4.1. Resonator theory

Before going straight to the design of the here presented LEKIDs, let us have a look at the theory of the two most common resonant circuits to get a better understanding of the properties of resonators in order to optimize them for our purposes. Fig. 4.1 shows these two types of resonators consisting of discrete elements. The circuit on the left is a parallel resonant circuit and on the right we see a series resonant circuit. The parallel resonator consists of an inductance L, a capacity C and a resistance R that takes the losses of the circuit into account. The applied voltage V is the same at each of these elements, whereas the current I is different in each element. The series resonator consists of the same discrete R, L and C with the difference, that the in this case the current is the same in all elements and the voltage varies for each of them. The impedance of these two circuits follows from *Kirchhoff's law* to

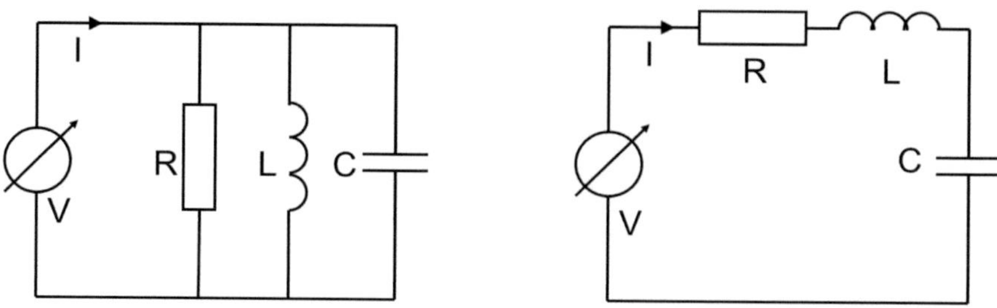

Fig. 4.1.: Schematic of a parallel resonant circuit (left) and of a series resonant circuit (rigth).

Parallel resonant circuit :

$$\underline{Z}_{in,p} = (\frac{1}{R} + \frac{1}{j\omega L} + j\omega C)^{-}1 \quad (4.1)$$

Series resonant circuit :

$$\underline{Z}_{in,s} = R + j\omega L + \frac{1}{j\omega C} \quad (4.2)$$

The resonance frequency f_0 for both resonator types is defined as the frequency, where the inductive reactance and the capacitive reactance reach the same value. In that case the imaginary part of the resonators impedance is zero. We can, therefore, derive the resonance frequency for both resonators by setting $Im\{Z_{in}\} = 0$ giving

$$\omega_0 = \frac{1}{\sqrt{LC}} \quad (4.3)$$

where ω_0 is the angular frequency with $\omega_0 = 2\pi f_0$. If a resonator is reaching its resonance frequency, it is able to store more and more energy. At its resonance frequency all energy, except the dissipated energy in R, is stored in same parts in the inductor and the capacitor or in other words it is stored either as magnetic energy or as electrical energy oscillating between them with the resonant frequency ω_0. Therefore the stored energy in a resonator can be calculated with

$$Stored\ energy\ in\ resonator:\ E_{res} = \frac{1}{2}CV^2 = \frac{1}{2}LI^2 \quad (4.4)$$

and the dissipated power in the resonator is simply defined as

$$Dissipated\ power\ in\ resonator:\ P_{res} = \frac{1}{2}\frac{V^2}{R} = \frac{1}{2}RI^2 \quad (4.5)$$

Resonators are characterized by their so-called quality factor Q. This dimensionless factor describes how much a resonator is damped due to ohmic losses and is defined as

$$Q = \omega_0 \frac{energy\ stored\ in\ resonator}{dissipated\ power\ in\ resonator} = \omega_0 \frac{E_{res}}{P_{res}} \tag{4.6}$$

With this definition we can calculate the quality factor for the parallel and the series resonator of Fig. 4.1 with

Parallel resonant circuit : *Series resonant circuit :*

$$Q_p = \omega_0 \frac{\frac{1}{2}CV^2}{\frac{1}{2}\frac{V^2}{R}} = \omega_0 RC = \frac{R}{\omega_0 L} \tag{4.7}$$

$$Q_s = \omega_0 \frac{\frac{1}{2}LI^2}{\frac{1}{2}RI^2} = \frac{\omega_0 L}{R} = \frac{1}{\omega_0 RC} \tag{4.8}$$

With these two main factors, the resonance frequency and the quality factor, the resonator can be characterized.

We will now consider a resonator coupled to an external circuit. As mentioned before, KIDs are used in multi-pixel arrays and are read out over a single feedline. Therefore the resonators are coupled to this transmission line, which adds an additional external impedance Z_{ext} to the circuits in Fig. 4.1. We will show the derivation for this network only for the series resonator case because it is the circuit that is used to model the LEKID resonators throughout this thesis. The equivalent circuit with the external impedance Z_{ext} is shown in Fig. 4.2. This external impedance changes the quality factor Q that we defined above, to the so-called loaded quality factor Q_L due to the ohmic loss of this impedance. From now on we define the quality factor as

$$\frac{1}{Q_L} = \frac{1}{Q_0} + \frac{1}{Q_{ext}} \tag{4.9}$$

where Q_L is the loaded quality factor, combining the unloaded quality factor Q_0 of the resonator and the quality factor of the external network Q_{ext}. As mentioned before the KIDs are coupled to a transmission line to excite the resonant circuit. The external network consists in our case only of this additional coupling impedance and we can therefore define $Q_{ext} = Q_C$. The coupling quality factor Q_C tells us

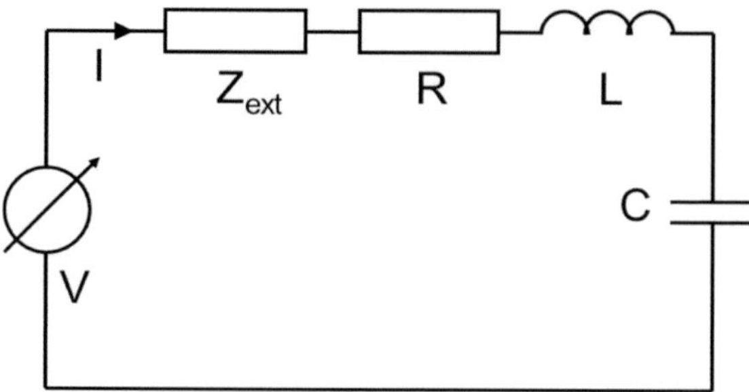

Fig. 4.2.: Schematic of a series resonant circuit coupled to an external network with impedance Z_{ext}.

how strongly the resonator is coupled to the transmission line. The quality factors can be derived from the measured scattering parameter S_{21}. In Fig. 4.3 a classical resonance curve of a two-port absorption resonator is shown. The accordant quality factors are defined as

$$Q_L = \frac{f_{res}}{\Delta f_L} \quad (4.10) \qquad Q_0 = \frac{f_{res}}{\Delta f_0} \quad (4.11)$$

where Δf_L is the classical half power bandwidth of a resonator to calculate Q_L and Δf_0 the corresponding bandwidth to calculate Q_0. Δf_0 and Δf_L define the frequency ranges within the energy is at least half its peak value, both for the pure resonator (Q_0) as well as the entire system (Q_L) respectively. The value, where the frequency bandwidths Δf_L and Δf_0 have to be taken depends on the depth of the resonance dip and can be calculated with [31, 11] to

$$S_{21,L} = \sqrt{\frac{1+|S_{21min}|^2}{2}} \quad (4.12) \qquad S_{21,0} = |S_{21min}|\sqrt{\frac{2}{1+|S_{21min}|^2}} \quad (4.13)$$

where $|S_{21min}|$ is the value of S_{21} at the resonance frequency of the resonator (see Fig. 4.3). For resonance dips with $|S_{21,min}| << 1$, where $1+|S_{21,min}| \approx 1$, Δf_L corresponds to the $-3dB-bandwidth$ as a good approximation. The depth of the resonance dip depends strongly on the coupling strength between the transmission line and the resonator. For a two port absorption resonator this coupling factor is

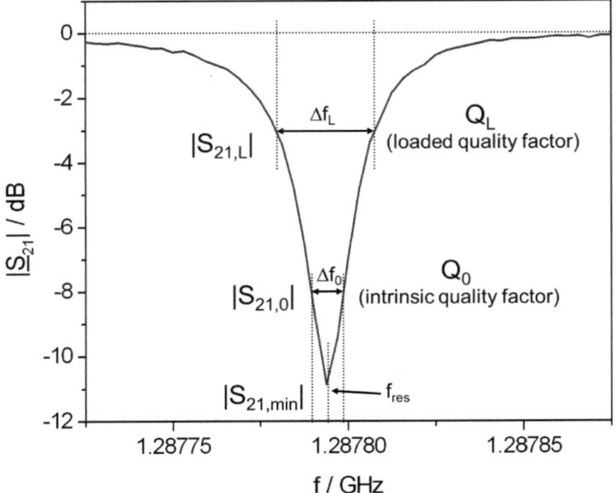

Fig. 4.3.: Resonance dip of a resonator coupled to a transmission line. The figure shows the determination of the loaded and unloaded quality factor from the measured scattering parameter S_{21}.

given by

$$\kappa = \frac{|S_{11,max}|}{|S_{21,min}|} = \frac{1 - |S_{21,min}|}{|S_{21,min}|} \tag{4.14}$$

where $|S_{11,max}|$ is the reflection scattering parameter at the resonance frequency. The quality factors are linked by this coupling factor κ that depends only on $S_{21,min}$. With κ as defined above, the quality factors can be calculated with

$$Q_0 = Q_L(1+\kappa) \tag{4.15}$$

$$Q_C = \frac{Q_0}{\kappa} \tag{4.16}$$

Therefore, the unloaded quality factor Q_0 and the coupling quality factor Q_C can directly be calculated from the loaded quality factor Q_L dependent of $S_{21,min}$ with

$$Q_0 = \frac{Q_L}{|S_{21,min}|} \tag{4.17}$$

$$Q_C = \frac{Q_L}{1 - |S_{21,min}|} \tag{4.18}$$

We see that for a high coupling factor ($|S_{21,min}| << 1$) Q_0 becomes very large depending on Q_L and $Q_C \approx Q_L$. The superconducting resonators presented in this work have very little losses and therefore high intrinsic quality factors Q_0. This allows designing the loaded quality factor for a resonator by optimizing the coupling factor. We also see from the equations above, that for a very weak coupling, where $|S_{21,min}| \approx 1$, $Q_C \to \infty$. In that case, the resonance curve is very narrow because of $Q_L \approx Q_0$. If Q_L is very high, the resonances can be packed closer, which is important for our application, due to a limited bandwidth of the readout electronics. As mentioned before, the response of a kinetic inductance detector is measured in resonance frequency, amplitude and phase of a resonant circuit. If Q_C is too high (coupling too weak), the resonance dip is too sharp leading to a saturation of the detector even for very small signals. For that reason, a compromise between high Q_L and sufficient coupling has to be taken into account for the design of a KID resonator, which will be discussed later.

4.2. The LEKID resonator circuit

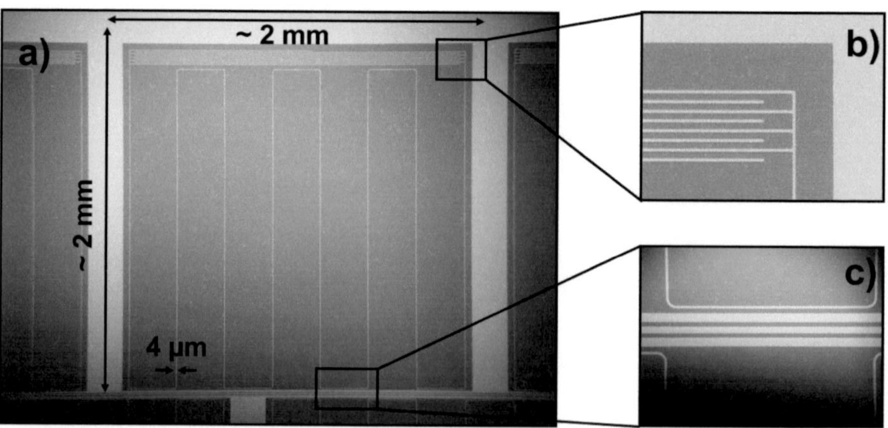

Fig. 4.4.: a) Picture of a lumped element kinetic inductance detector coupled to a coplanar wave guide on a silicon substrate. b) enlarged part of the interdigital capacitor; c) zoom on the coupling area of the coplanar waveguide.

The design of the LEKIDs consists of a long meandered line, the inductive part, and an interdigital capacitor both building the resonant circuit. In Fig 4.4 one of these LEKIDs is shown. It is coupled inductively to a coplanar waveguide (cpw) transmission line and surrounded by a ground plane. A cpw-line is chosen for optical coupling reasons. The ground plane on the substrate bottom (in the case of a microstrip line) would make the optical coupling impossible the way it is done for the LEKIDs presented here (see Chapter 7). The LEKID combines the resonator circuit with a direct absorption area, which can be optimized over the width and the spacing between the meander lines and the normal resistivity of the superconducting material, which will be discussed later in detail. Fig 4.5 shows an equivalent circuit scheme of the LEKID coupled inductively over a cpw-line. The resonator itself is defined by the capacitance C, the inductance L, the resistance R and the mutual inductance M. The coupling is caused by magnetic flux from the current in the cpw-line threading the meander section. The mutual inductance M is defined as magnetic flux per current flowing in the transmission line:

$$M = \frac{\partial \phi}{\partial I} \tag{4.19}$$

M depends therefore on the distance between the meander section and the cpw-line. In Fig. 4.6a), the current density distribution $\vec{j}$ and the corresponding magnetic field distribution $\vec{H}$ of a cpw-line with finite ground plane is shown. The current density

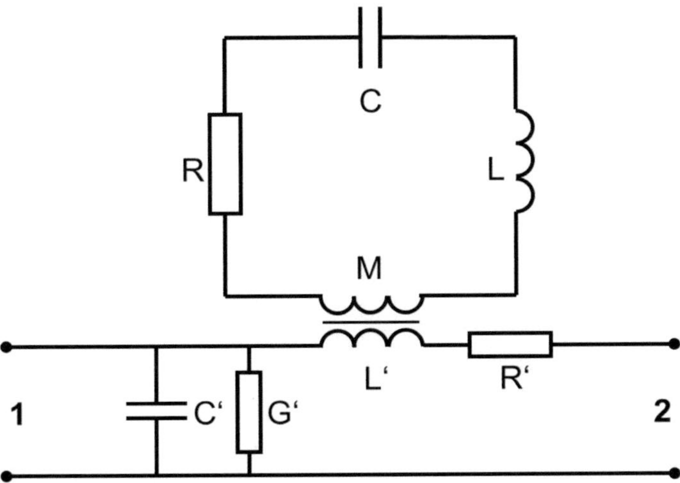

Fig. 4.5.: Equivalent circuit of a LEKID resonator inductively coupled to a transmission line via a mutual inductance M.

has its maximum at the edges of the center line and at the ground plane edges next to the center line. In the ground plane it decreases exponentially to zero within a distance of around $5 \cdot \mathrm{w1}$ [49]. By this, the magnetic field surrounds the center line and the ground plane. The field intensity is hereby higher next to the center line compared to the outer edges.

To couple now a LEKID to such a cpw-line, the ground plane has to be narrowed down where the resonator is coupled, to increase the magnetic flux threatening the meander section and therefore the coupling strength. This is demonstrated in Fig. 4.6b), where m indicates the width of the ground plane and w the width of the meander line section that is parallel to the transmission line. The further the resonator from the cpw-line (defined by m and the distance between ground plane and resonator), the weaker the magnetic field threatening the meander and the weaker the resonator is coupled to the cpw-line. A more detailed derivation of M in the case of a microstrip transmission line can be found in [16].

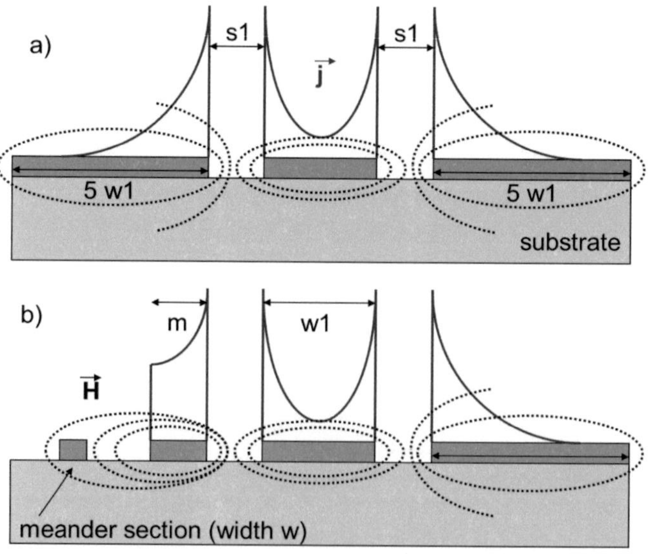

Fig. 4.6.: a) current density ($\vec{j}$) and magnetic field ($\vec{H}$) distribution of a cpw-line with finite ground plane. b) current density and magnetic field distribution of cpw-line when a LEKID is coupled to the transmission line.

The characteristic impedance of a cpw transmission line is in general given by

$$\underline{Z}_L = \sqrt{\frac{R' + j\omega L'}{G' + j\omega C'}} \tag{4.20}$$

where R', G', L' and C' are the resistance of the conductor, the conductance of the dielectric, the inductance and the capacitance per unit length. For a superconducting cpw-line, where the ohmic losses are very small, Z_L can be simplified to

$$Z_L = \sqrt{\frac{L'}{C'}} \tag{4.21}$$

Generally the external networks such as connectors, cables and network analyzer outputs are matched to 50 Ω. To avoid reflections and standing waves due to impedance jumps, the cpw-line, in this case, is matched to $Z_L = 50\ \Omega$, which can be defined by the dielectric constant of the substrate and the ratio of the line width and the gap width [49]. The transmission line dimensions used here, are $w_1 = 20\ \mu m$ and $s1 = 12\ \mu m$. This geometry is chosen as a compromise between coupling strength, array filling-factor and microfabrication limits.

4.2.1. Modelling a LEKID resonator for the NIKA specifications

The resonance frequency of the LEKID circuit is simply given by

$$f_{res} = \frac{1}{\sqrt{LC}} \tag{4.22}$$

where L is the total inductance of the meander line and C the capacitance of the interdigital capacitor. L consists in our case of the geometrical inductance L_{geo} (see 3.2) and the kinetic inductance L_{kin} ($L = L_{geo} + L_{kin}$), where

$$L_{geo} = L_{m,int} + L_{m,ext} + M = L_{line} + M \tag{4.23}$$

The inductive part M of the coupling impedance is included in the geometrical inductance ($L_{geo} = L_{line} + M$) and is in general much smaller than the inductance of

the meander line L_{line}. To determine the resonance frequency, L_{line} and C have to be designed and calculated. f_{res} should hereby be between 1 and 2 GHz corresponding to the bandwidth of the cryogenic amplifier that is used for later measurements. Another reason for a preference of low resonance frequencies is the increase of surface and radiation losses that are lowering the unloaded quality factor.
To design a LEKID geometry that absorbs photons with a wavelength of 2.05 mm, the pixel size has to be chosen in this range for practical optical reasons that will be explained later. Therefore the length of the meander line is more or less defined by the size of the pixel. The spacing between the meander lines and their width has to be calculated for an optimal optical coupling and define the number of meander lines and thus the inductance L_{line}. In addition, to increase the kinetic inductance, the cross section of the meander line should be kept to a minimum. Here we use line widths between 3 and 4 μm and film thicknesses in the range of d=30 nm giving kinetic inductance values, at an assumed temperature of $T = 200\ mK$, of around $L_{kin} = 1.2\ pH/\square$ for aluminum and $L_{kin} = 0.3\ pH/\square$ for niobium respectively. This means, that there are few options to change the resonance frequency by the inductive part in order to guarantee an identical direct absorption area for all pixels. For fixed meander line geometry, f_{res} is therefore tuned over the length and number of fingers of the interdigital capacitor. The current density in this capacitor is relatively small compared to the inductive line and a modification in geometry has therefore negligible influence on the photon absorption properties of the LEKID and thus for the responsivity.
To determine the absolute value of L_{line} and C let us consider them separately. In other words, we simulate the two components as a serial impedance in a two port system as shown in Fig. 4.7. Z represents the impedance of either the inductan-

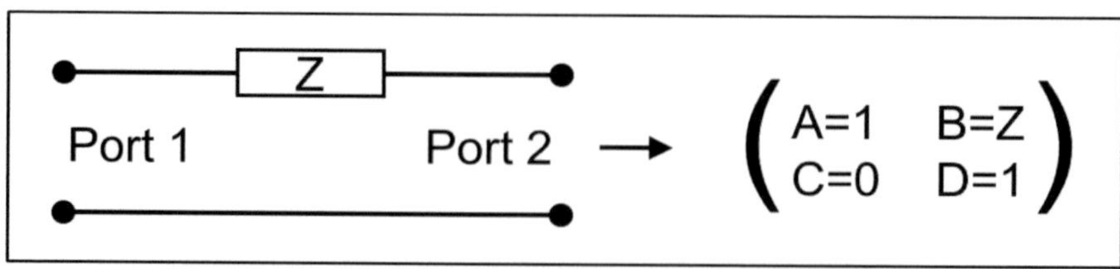

Fig. 4.7.: Transmission line model of a series impedance in a two port system with corresponding ABCD-matrix.

ce or the capacitance. Two port systems are usually characterized by its scattering parameters S_{11}, S_{12}, S_{21} and S_{22}. These S-parameters can be described by a ABCD-matrix [49] for various circuit configurations. In this case, for the series impedance, the matrix is defined as shown in Fig. 4.7, where we see that B corresponds to the impedance Z and can therefore be expressed by the S-parameters of the system given by

$$B = Z = Z_0 \frac{(1+S_{11})(1+S_{22}) - S_{12}S_{21}}{2S_{21}} \tag{4.24}$$

The S-parameters of the series inductance and capacitance can be extracted, for a example, from a *Sonnet* simulation [57], which offers the possibility to simulate superconducting properties at high frequencies and is therefore a suitable simulation tool for this application. This means, that we can simulate our resonator geometries including the kinetic inductance L_{kin}. The surface impedance at a given temperature can hereby be defined by the surface resistance R_S and the surface inductance L_S, which corresponds to L_{kin} and can be calculated with (3.10). The corresponding *Sonnet* geometries for L_{line} and C are shown in Fig. 4.8a) and b). A method to calculate L_{line} and C analytically for a first estimation, before designing a simulation model, is described in the appendix A.2. From the simulated S-parameters and Z_0 (port impedance), Z can be calculated using 4.24. Taking the imaginary part of Z, the values for L_{line} and C can be derived with

$$L_{line} = \frac{Im(Z)}{\omega} \tag{4.25}$$

$$C = \frac{-1}{Im(Z)\omega} \tag{4.26}$$

Let us first consider a resonator design consisting of a meander line as shown in Fig. 4.8a) combined with the interdigital capacitor of Fig. 4.8b) and calculate the absolute values of L_{line} and C. For a Nb resonator with a 60 nm thick film, assuming a temperature of around $T = 4\ K$, we can calculate the kinetic inductance using the theory in Chapter 3. Table 4.1 shows the geometrical and electrical parameters used in the simulation to determine the inductance L_{line}, the capacity of the interdigital capacitor C and the resonance frequency f_{res}. Taking the coupling to the transmission line not into account, we calculate a resonance frequency of $f_{res} \approx 1.27\ GHz$, which is well placed in the cryogenic amplifier bandwidth.

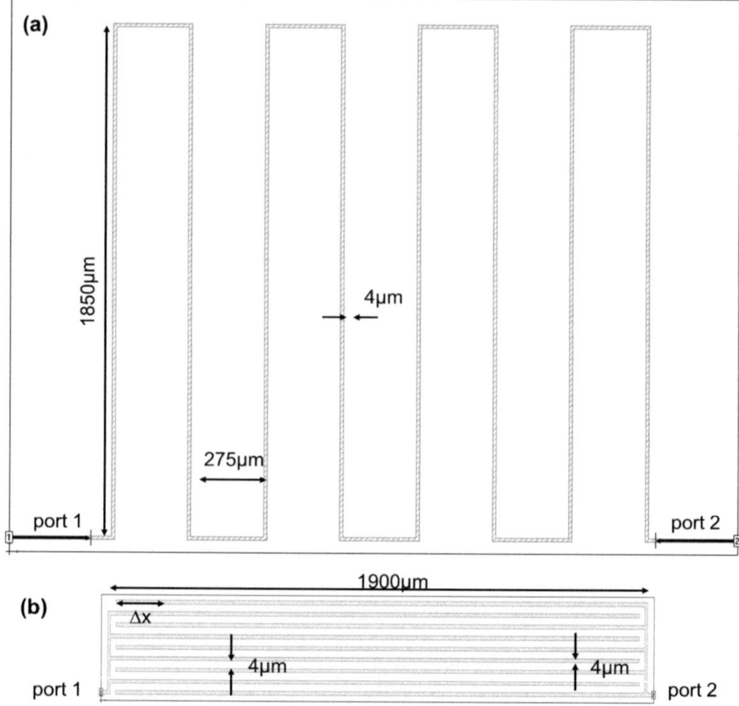

Fig. 4.8.: *Sonnet* model of a typical LEKID inductive meander line (a) and interdigital capacitor (b) to determine L_{line} and C.

If we now couple the resonator to a transmission line as shown in Fig 4.9, we can simulate the resonance frequency f_{res+M} including M and calculate it. From the difference in resonance frequency $(f_{res} - f_{res+M})$, M is given by

$$M = \frac{1}{\omega_{res+M}^2 C} - L_{line} \tag{4.27}$$

where ω_{res+M} is the angular resonance frequency of the circuit including M (see Fig 4.9). For a coupling with $Q_C \approx 30000$ and $f_{res+M} = 1.163 GHz$, we calculate for the coupling inductance $M \approx 3\ nH$. This depends of course on the coupling strength, but we can see, that M is much smaller than L_{line} and has therefore less but not negligible influence on the resonance frequency of the resonator. The ratio of L_{line} and M depends hereby on the line inductance, determined by the cross section and the length of the line, and M. By using short and wide lines for the inductive part

material	niobium
T_C	9.2 K
meander line width	4 μm
meander line length	16.7 mm
capacitor line width	4 μm
capacitor line spacing	4 μm
capacitor line number	9
capacitor line length	1.9 mm
R_s	$5 \cdot 10^{-8}\ \Omega/\square$
L_{kin}	0.2 $pH/\square$
assumed temperature	4.2 K
L_{line}	19 nH
C	0.82 pF
f_{res}	1.27 GHz

Table 4.1.: Parameters of one possible Nb LEKID pixel design to determine L_{line}, C and the resonance frequency f_{res}. The results have been calculated using the S-parameters simulated with *Sonnet*, where R_s and L_{kin} are derived as described in Chapter 3

and a relatively strong coupling, it is possible to achieve values for M that exceed the line inductance. For the LEKID geometries we deal with throughout this thesis, the line inductance L_{line} is always higher than the mutual coupling inductance M and the resonance frequency is dominated by the capacitor C and L_{line}.

This method, to determine the resonance frequency, gives a good estimation and can be used as simulation tool to design the LEKID for a given resonance frequency by the geometry of the meander line and the interdigital capacitor. A more detailed characterization of the electrical coupling to the transmission line, quality factors and the frequency tuning is discussed in Chapter 6.

The LEKID geometry shown above consists of a very narrow line of only 4 μm, which demands a fabrication process that guarantees high quality films and lithography/etching methods. High frequency losses in such narrow and thin films are dominated by surface scattering effects and depend therefore on the homogeneity of the film and the quality of the structure edges where the etching takes place. In the following chapter different methods of film deposition and microprocessing methods to fabricate the LEKID structures are demonstrated and discussed.

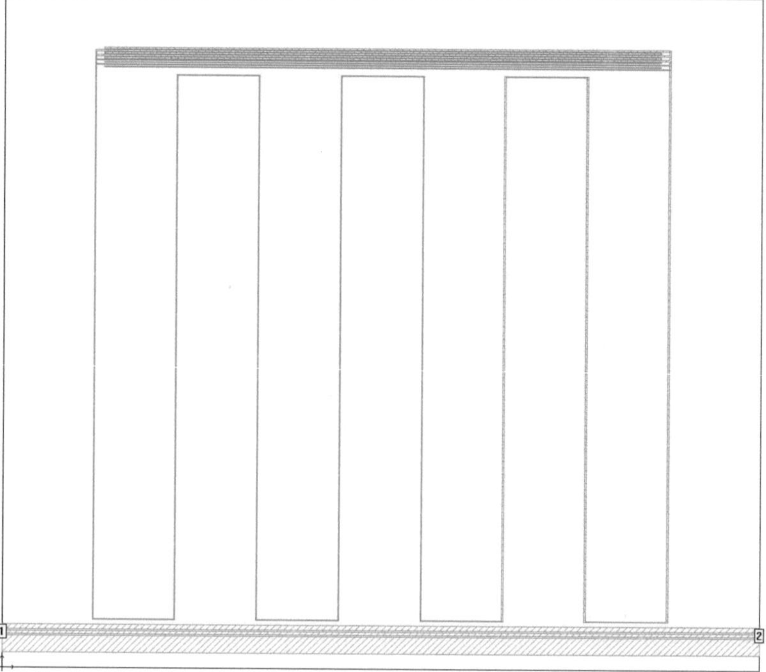

Fig. 4.9.: *Sonnet* model of a typical LEKID inductively coupled to a coplanar transmission line.

5. Film deposition and fabrication process

The performance of a kinetic inductance detector not only depends on the design but also on the quality and properties of the superconducting material. As we have seen from the superconducting theory, the normal conductivity of the superconductor has non-negligible influence on the high frequency surface losses and on the kinetic inductance.

From chapter 3 we know, that the kinetic inductance L_{kin} increases with lower film volume and lower Cooper-pair density n_s, but the surface losses in Z_s also increase. If these losses are too high, the quality factor of the microresonator decreases and lowers the sensitivity of the detector. To increase the kinetic inductance of a superconducting film with a relatively low normal conducting charge carrier density n_n, we need to use thin films in the range of d=20-40 nm. The London penetration depth λ_L is then much higher than the thickness of the film, leading to a homogeneous current density over the whole cross section of the conductor and therefore to a much higher kinetic inductance (see chapter 3). The quality of very thin superconducting films has thus to be optimized to achieve sufficient high kinetic inductance and films with homogeneous thickness over the entire wafer surface. For that reason, the deposition and fabrication of the superconducting circuits play a major role in terms of sensitivity of the KIDs.

In Fig. 5.1 a general process schematic of the LEKID fabrication is shown. The superconducting film is deposited either via sputtering or evaporation on a 2 inch high resistivity (>5 kΩcm) silicon substrate. After the film deposition, a positive photo resist is spun on top of the metallization for the following lithography step, which is done using a chrome mask and an UV-aligning system. After the development of the photo resist, the films are etched either by using a wet etch process or a reactive ion etching (RIE) process depending on the material. This fabrication process is used for all LEKID structures shown in this work. In the following we present the fabrication process more detailed for the two materials niobium and aluminum, which are the main materials used here. The thin films are characterized

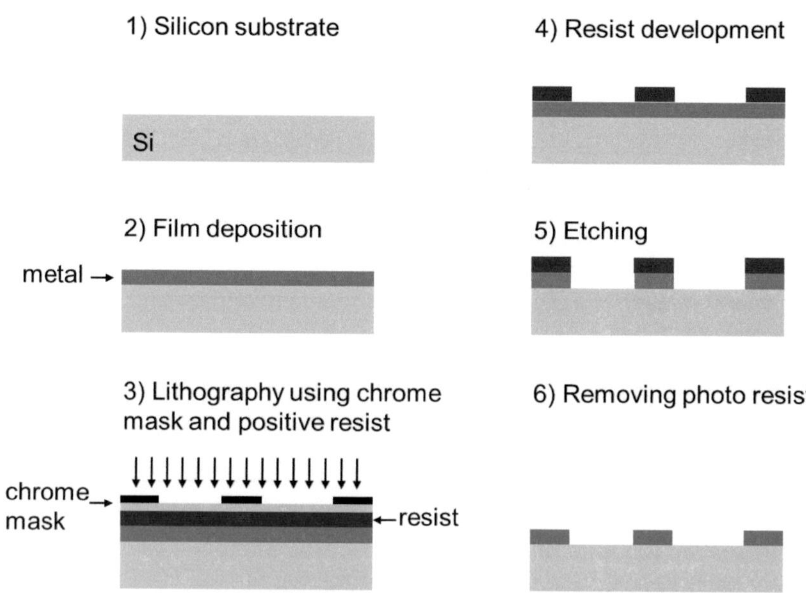

Fig. 5.1.: Process schematic of the LEKID fabrication.

by low temperature measurements regarding the resistivity depending on the deposition process and the film thickness. Also the microprocessing of the films will be discussed, such as lithography and etching processes for the two materials.

5.1. Deposition of niobium and aluminum

Niobium (Nb) For the first tests and measurements to characterize KIDs, Nb was used. Niobium is a type II superconductor with a critical temperature of 9.2K [60]. This allows measurements in liquid helium at $T = 4.2\ K$, which is less time consuming than materials with a $T_C < 4.2\ K$. The niobium used for the KIDs is deposited via dc-down-sputtering on a 300 μm thick high resistivity silicon substrate with $R > 5\ k\Omega cm$. A schematic of the sputtering principle is illustrated in Fig. 5.1. The sputtering gas pressure, in our case argon, between the target (solid niobium) and the substrate in the vacuum chamber gets ionized due to the applied dc-power. The argon ions are therefore accelerated to the target and break niobium atoms out of it. Further, these atoms diffuse in the chamber until a part of them is deposited on the

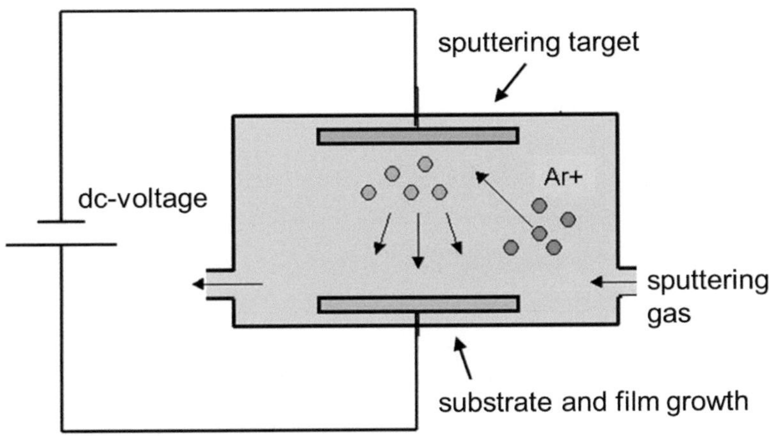

Fig. 5.2.: Schematic of a dc-sputtering system using an argon plasma. The argon ions break atoms out of the target material, which are deposited on the substrate.

substrate and forms the niobium film. The sputter velocity or sputtering rate depends on the applied dc-voltage, the amount of argon gas, the pressure in the chamber and the distance between the target and the substrate. The vacuum pressure in the chamber, before adding the Argon gas, is $p \approx 5 \cdot 10^{-8}$ mbar.

The niobium we used is deposited with a dc-power of $P = 600\ W$, an argon flow of 50 sccm and a process pressure of 1.3 Pa giving a sputtering rate of 3 nm/s. With this process we measured for a 60nm thick niobium film a $T_C = 9.2\ K$ and a square resistance of around 0.8 Ω at $T = 10\ K$ and 4 Ω at room temperature (300 K). A parameter that is used for the characterization of thin metal films, is the *Residual Resistance Ratio* (*RRR*). It is defined as the ratio of the dc-resistance of the film at $T = 300\ K$ divided by the resistance just above the critical temperature T_C. The higher this ratio the higher is the purity of the film. For our 60 nm thick niobium film we find $RRR \approx 5$. This value depends also strongly on the thickness of the film, where the RRR decreases rapidly for very thin films ($d < 100\ nm$).

Aluminum (Al) For the aluminum films we used two different deposition methods. One by dc-sputtering, as shown above for niobium, and a second one using an electron beam evaporation system. The principle of e-beam evaporation is shown in Fig. 5.2. An high energy electron beam heats the target, in this case aluminum,

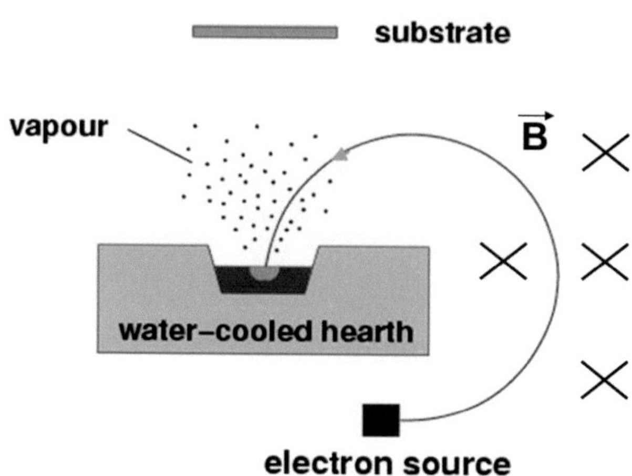

Fig. 5.3.: Schematic of evaporation system using an high energy electron beam to heat and evaporate the target material.

hold in a small spot until it gets liquefied and later evaporates on the substrate, which is situated above. This deposition method is much more directive than the sputtering process. This is due to the lower process pressure in the chamber, which is the same as the vacuum pressure of around $p = 5 \cdot 10^{-8}$ mbar. The deposition velocity can be controlled by the power of the electron beam. For our aluminum we used a deposition rate of 1 nm/s, which was found, by varying the beam power, to give the best film quality. The evaporation process is known to achieve better film quality due to a better stoichiometric growth, but it is more difficult to control the deposition ratio. The atmosphere in the chamber is cleaner due to the absence of Argon gas, as it would be the case for a sputter process, leading to a much lower vacuum pressure during the process. The difficulty of controlling the deposition rate is especially for aluminum the case because it has a sharp transition from the liquid to the vapor state, it sublimates immediately.

For the sputtered aluminum we achieved the best film quality by using a dc-power of $P = 300\ W$, an Argon flow of 42 sccm and a process pressure of $p = 1.3\ Pa$. With this process we get a deposition rate of 1.3 nm/s. Table 5.1 shows the square resistances and the RRR value for different thicknesses of niobium and aluminum films, deposited with the described processes. We see from table 5.1 that the film quality, regarding the RRR, is different for the sputtered and the evaporated alu-

material	deposition	thickness	$R_\square$ (300K)	$R_\square$(4.2K)	RRR
Nb	sputtering	60 nm	4 Ω/□	0.8 Ω/□	5
Al	sputtering	40 nm	1.4 Ω/□	0.5 Ω/□	2.8
Al	sputtering	20 nm	3.6 Ω/□	1.8 Ω/□	2
Al	evaporation	92 nm	0.4 Ω/□	0.07 Ω/□	5.7
Al	evaporation	65 nm	0.8 Ω/□	0.21 Ω/□	3.81
Al	evaporation	40 nm	1.21 Ω/□	0.33 Ω/□	3.66
Al	evaporation	16 nm	5.95 Ω/□	3.65 Ω/□	1.63

Table 5.1.: Comparison of material parameters using different deposition methods for niobium and aluminum. The RRR is calculated from the square resistances $R_\square$ at 300 K and 4.2 K.

minum for the same thickness. The resistivity at room temperature and at 4.2 K is lower for the evaporated film due to the higher purity. The square resistance and so the thickness of the film plays a major role for the detectivity of the LEKIDs, not only because the kinetic inductance fraction is higher due to the smaller volume but also the optical coupling of a mm-wave signal to the detector depends directly on the conductivity of the material (see later in chapter 7).

The aluminum that showed better detectivity is the sputtered aluminum in spite of the lower conductivity and RRR. The exact reasons for this behavior have not been studied in the framework of this thesis and needs therefore further investigations in future work. Due to the obtained results, we decided to use the sputtered aluminum films for the LEKIDs. Therefore, a series of aluminum film sputter depositions with different thicknesses have been performed and the films have been characterized at low temperatures. Fig. 5.4 shows the measured conductivity in MS/m for different thicknesses of sputtered aluminum films at room temperature and at 4.2 K. It demonstrates, that the conductivity is not constant, but decreases for decreasing film thickness. The difference at room temperature in conductivity between a 120 nm thick aluminum film and a 10 nm thick film is around a factor of two, whereas at low temperatures we find a factor of around 7. This effect is important regarding the calculation of the kinetic inductance using the $Mattis-Bardeen\ equations$ introduced in chapter 3.3, which depends directly on the conductivity of the superconducting film (σ_n). From the conductivity at room temperature and at 4.2 K, the residual resistance ratio (RRR) has been calculated for the different sputtered aluminum films. The results are also shown in Fig. 5.4. As the conductivity, the RRR

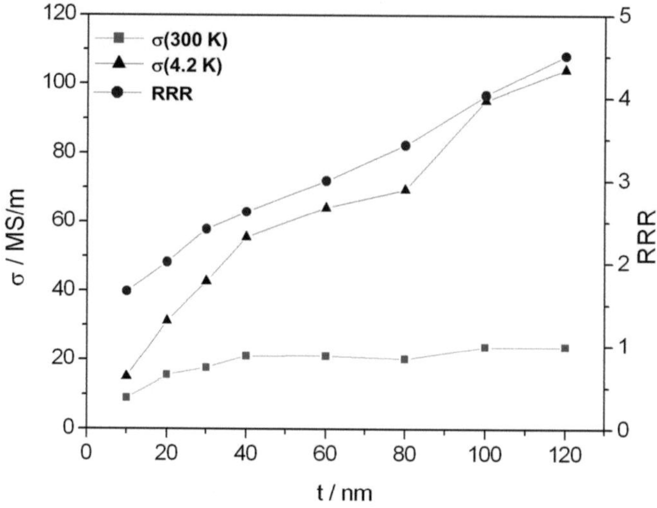

Fig. 5.4.: Conductivity of sputtered aluminum for different film thicknesses at room temperature (red line) and at 4.2 K (black line) with corresponding residual resistance ratio (RRR) (blue line).

decreases also with decreasing film thickness, which means that the RRR of the aluminum depends not only on the deposition method but also on the thickness of the film. The RRR varies with film thickness from around 1.6 for a 10 nm film to around 4.5 for a 120 nm thick film.

To demonstrate the influence of the film thickness on the kinetic inductance of thin aluminum films, we calculated L_{kin} as shown in chapter 3.3 for the characterized aluminum films presented above. For the aluminum, we calculate the gap energy, using the approximation $2\Delta = 3.526 k_B T_C$, to around 0.182 eV at $T = 0\ K$. We can now determine the complex conductivity $\sigma = \sigma_1 - j\sigma_2$ with the measured conductivity σ_n shown in Fig. 5.4. From the complex conductivity σ and the film thickness we can then calculate the kinetic inductance using equation (3.15). The results are presented in Fig. 5.5 for different aluminum film thicknesses. We see that the kinetic inductance increases exponentially with decreasing film thickness according to the theoretical model. We will therefore use films that do not exceed a thickness of d=40 nm to increase the sensitivity of our detectors. The results also show the importance of a homogeneous film thickness over the surface of an entire silicon wafer regar-

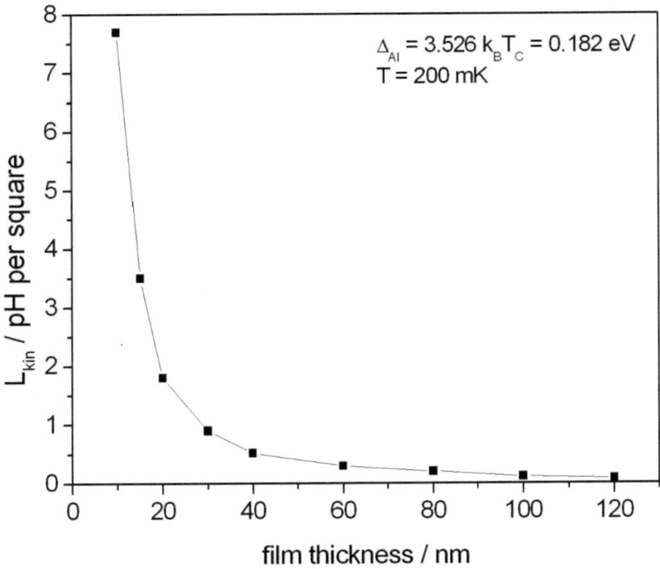

Fig. 5.5.: Calculation of the kinetic inductance of aluminum films depending on the film thickness at an assumed temperature of $T = 200\ mK$.

ding an array of several hundreds of pixels. If the film thickness varies too much due to a non-optimized deposition process, the designed resonance frequencies on the array can vary due to a different absolute value of the line inductance L_{line} shown before. The variation of film thickness on a 2 inch wafer, comparing the thickness at the center and the edge of the wafer, was measured to be around 5%, which is small enough to avoid significant resonance frequency shifts for superconducting aluminum resonators.

5.2. Microprocessing of niobium and aluminum films

For thin superconducting films, whose thickness is smaller than the London penetration depth λ_L, the main microwave losses in the resonator are caused by surface roughness effects [68]. To reduce these effects, not only the top side of the films has to be clean and homogeneous but also the surface at the etched borders of the structures, which demands a well optimized lithography and etching process for the KIDs.

Both, the aluminum and the niobium films are structured by UV-lithography using a chrome mask and a positive photo resist. We decided to use an etching process instead of an lift-off process to guarantee better edge qualities of the fine and relatively long meander line of the LEKID structure.

Niobium The niobium films are etched by reactive ion etching (RIE) using SF_6 as reactive gas. The problem with this etching process is, that it etches also into the silicon substrate. It is therefore important to know the exact etching rate of the process in order to avoid uncontrollable etching into the silicon. If too much of the silicon around the resonator structure is etched away, this can change the effective $\varepsilon_{r,eff}$ and lead to a change in resonance frequency of the LEKID. Fig 5.6 shows a 60 nm niobium film on a silicon substrate after the etching process. On the left hand side we can see a part of the interdigital capacitor of the LEKID. A zoom on the border of the line is presented on the right hand side of the picture. It shows that the process is sufficient directive, there is almost no under etching of the structures visible. In this case around 500 nm of the silicon has been etched to show the quality of the process. Usually the etching into the silicon is between 50 and 100 nm. Also the borders of the niobium line appear very clean and sharp, which limits surface losses of the material at high frequencies to a minimum. The etching rate that was used for the niobium is 2 nm/s.

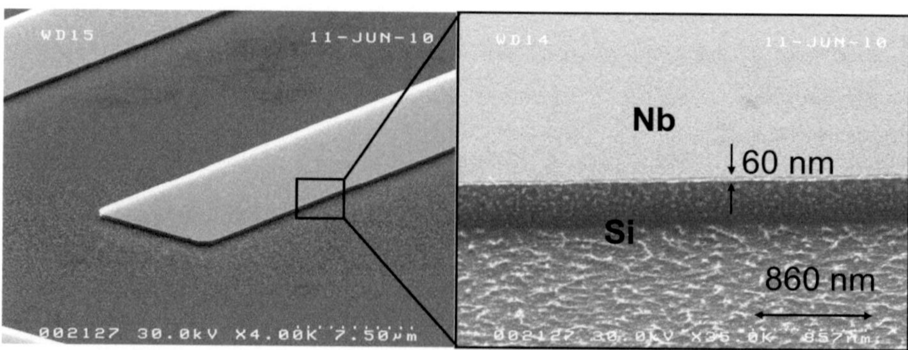

Fig. 5.6.: Niobium LEKID structure after reactive ion etching. left hand side: zoom on the interdigital capacitor; right hand side: high zoom on the border of a thin niobium line.

Aluminum For the aluminum film, one possibility is to etch it with a process that uses chlorine. This process is not being used in our case for security standard reasons in the IRAM clean room. We decided therefore to use a wet etching process to structure the aluminum films. One advantage of this wet etch method is that it does not etch into the silicon substrate and it can thus be used as an etch stop. Due to the thin films of only d = 20 - 40 nm, under-etching of the structures does not cause any problems here. To have a better look on the borders of the etched structures, we etched after the aluminum etching into the silicon with the process used for niobium. This is shown in Fig 5.7. On the left hand side two wide zoom pictures are shown and on the right hand side we see two pictures with a higher zoom on the borders of the structures. Also the aluminum wet etching process shows very clean structures and is comparable to the results of the niobium RIE process.

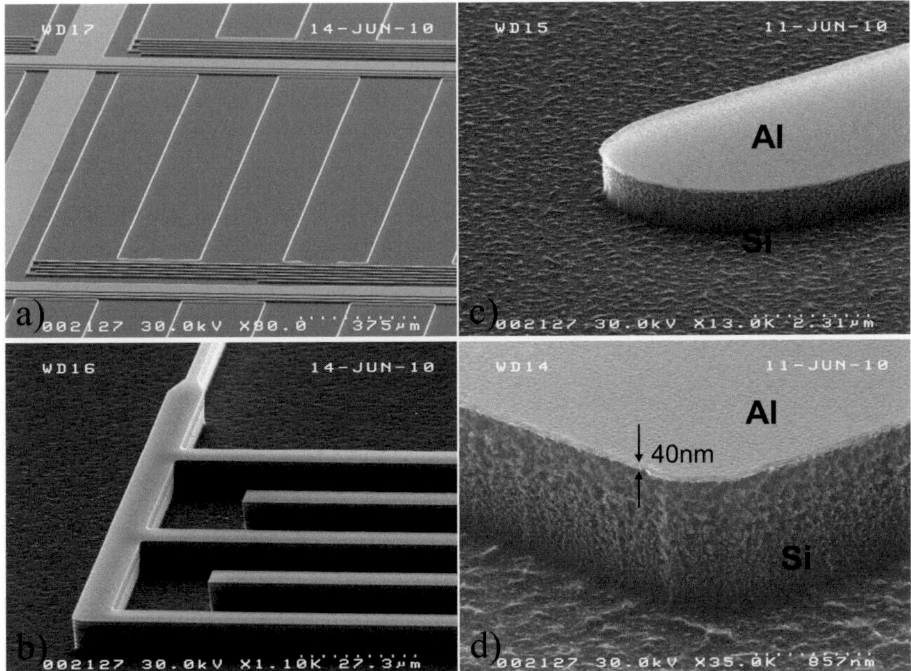

Fig. 5.7.: Aluminum LEKID structures on silicon substrate after wet etching of the aluminum and etching of silicon. a) and b) wide zoom on different parts of the LEKID; c) and d) high zoom to the borders of the fine aluminum lines.

Based on these results, simulations and measurements of niobium and aluminum LEKID multi pixel arrays have been done to investigate the electrical properties such as the coupling to the transmission line, quality factors, the resonance frequency tuning and the electromagnetic cross coupling between pixels. Simulations and measurements concerning these parameters are presented and discussed in the next chapter.

6. Investigation of electrical properties of multi-pixel LEKID arrays

To build large LEKID arrays, parameters such as coupling factor, quality factor and resonance frequency have to be adapted and optimized for the NIKA specifications. For that reason, at the beginning, arrays of six pixels have been designed and fabricated to measure the resonators at low temperatures and to investigate the mentioned parameters. The design of one of these six pixel arrays is shown in Fig. 6.1. It is a 10x10 mm sample, where the six pixels are coupled from both sides to the $cpw-line$, which is matched to an impedance of $Z_L = 50\ \Omega$. At the edge of the sample a so-called taper structure was added in order to have a larger area to bond from the sample holder to the sample and to keep the 50 Ω unchanged to avoid reflections due to an impedance mismatch. For these first tests of LEKID arrays we decided to

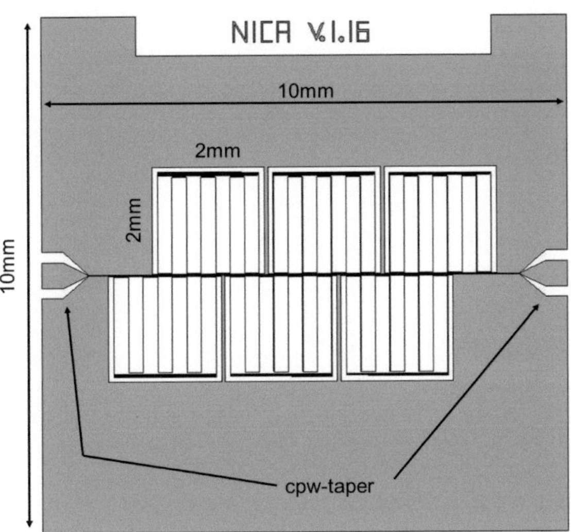

Fig. 6.1.: 6 pixel LEKID array.

use niobium as superconducting film. Niobium allows measurements in liquid helium (4.2 K) due to its T_C at 9.2 K. This is less time consuming to cool down the sample below its critical temperature than it would be for aluminum ($T_C = 1.2\ K$). Since we are investigating the electrical properties of the LEKIDs, the difference of the kinetic inductance fraction between aluminum and niobium (factor 10) does play a minor role. The higher kinetic inductance of aluminum becomes more important when we characterize the LEKIDs optically. The niobium is deposited and structured as described in Chapter 5. The used substrate for these measurements is a high resistivity ($> 5\ k\Omega cm$) silicon substrate with a dielectric constant of $\varepsilon_r = 11.9$. The samples have been measured in a cryostat with a base temperature of 2 K. This was achieved by pumping the helium chamber, leading to a decrease in temperature of the liquid helium.

In this chapter several parameters will be investigated, concerning the coupling, the frequency tuning and the quality factors. These parameters are defined by the geometry of the LEKID and its surrounding environment, consistent of the cpw-line and a ground plane frame. Fig. 6.2 shows a zoom on the coupling area of two neighboring LEKIDs coupled to a transmission line, where the geometrical dimensions are defined. w1 and $s1$ define the cross-section of the coplanar transmission line and therefore the impedance of 50 Ω. m and d define the coupling distances between LEKID and transmission line and g and a the geometrical spacing between

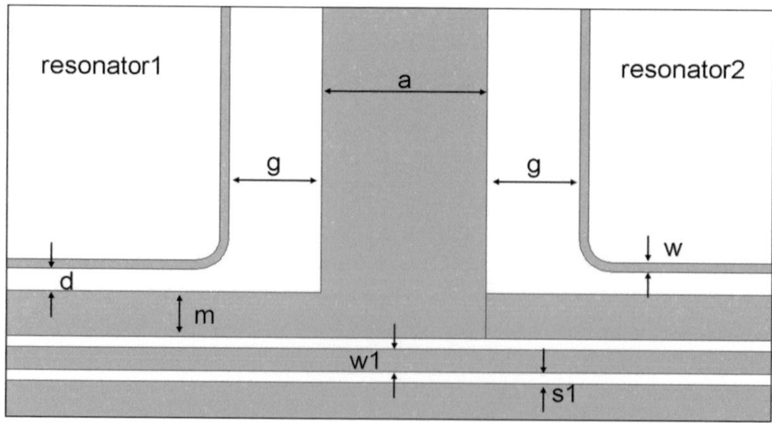

Fig. 6.2.: Definition of the dimensions of the characteristic LEKID geometry parameters determining the coupling and geometrical spacing between resonators.

neighboring LEKIDs. The influence of these geometrical dimensions has to be understood and optimized in order to design large arrays of highly sensitive LEKIDs. We will keep these definitions throughout the following chapters.

6.1. Coupling to the transmission line

The first parameter to optimize is the coupling factor κ and the coupling quality factor Q_C, which limits how much power from the cpw-line is coupled into the resonator and defines the shape of the resonance curve. As we have seen before, $Q_C \approx Q_L$ for $Q_0 >> Q_C$ (see 4.1). For our superconducting microresonators this is usually the case due to a very small surface resistance R_s allowing to chose Q_L by designing the coupling strength accordingly. A strong coupling (κ>1) gives deep resonance dips and a low Q_C and Q_L (wide shaped resonance curve). A weak coupling ($\kappa < 1$) on the other hand leads to a high Q_C and Q_L (narrow resonance curve), but also to less deep resonances. Therefore, κ is also a limiting factor for the sensitivity of the LEKID and for the packaging of a large number of resonators in a limited bandwidth. If Q_L is too low, the bandwidth of the resonance is too wide, which can lead to an overlap of the resonance curves of two neighboring resonators. If κ is too small, the resonance dip is not deep enough and Q_L very high, which limits the dynamic range of the LEKIDs and saturates the detector even for very weak signals. Therefore a tradeoff between a sufficient coupling strength and an acceptable Q_L has to be found. With the defined coupling factor κ of Chapter 4.1 and [16], an expression for the transmission scattering parameter S_{21} for a fixed Q_0 can be derived to

$$S_{21} = \frac{2}{2 + \frac{2\kappa}{1+j2Q_0\delta}} \tag{6.1}$$

where δ is the double sided fractional bandwidth defined as

$$\delta = \frac{2\Delta\omega}{\omega_0} = \frac{2(\omega_0 - \omega)}{\omega_0} \tag{6.2}$$

This is only valid for frequencies near resonance ($\Delta\omega << 2\omega$). In Fig. 6.3 calculations of S_{21} for three different values of κ are shown. Q_0 is set to 100000 for each

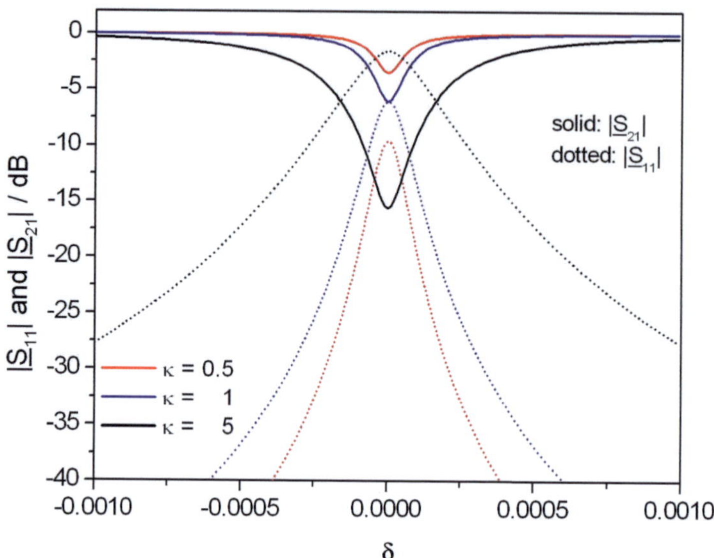

Fig. 6.3.: Calculation of S_{11} and S_{21} of an absorption resonator coupled to a transmission line for three different coupling factors κ.

curve and κ has the values 0.5, 1 and 5. From the definition of κ we see, that for $\kappa = 1$, S_{11} and S_{21} have the same value. In this case the resonance dip shows a depth of $|S_{21}| = -6\ dB$ and is called the critical coupling. From the calculation we can also see the influence of the coupling strength to the sensitivity and dynamic range of the detector. As we know, the resonance frequency shifts downwards and the amplitude of S_{21} increases if photons with energy $E > 2\Delta$ break Cooper pairs into quasi-particles. If the resonance dip is too narrow and too small, it approaches the carrier amplitude of the transmission line (in Fig. 6.3 at 0 dB) and disappears even for very small signals. This of course would be a highly sensitive detector device, but limited to very small variations in power that can be detected, leading to a small dynamic range.

This principle of the dynamic range dependency on the coupling factor is illustrated and explained in Fig. 6.4 on the example of two resonators with different couplings in the S_{21}- and IQ-plane. The blue line represents a resonator with a strong coupling, whereas the red line shows a measurement of a weakly coupled resonator. To

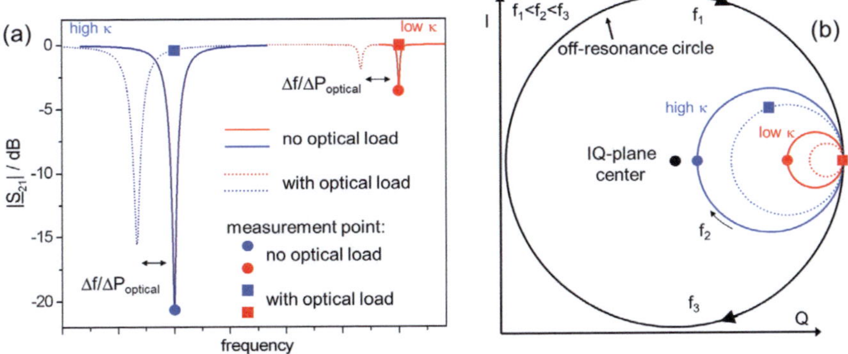

Fig. 6.4.: Dynamic detector range limited by the coupling strength: frequency response of two resonators with different coupling strength κ in S_{21}-plane (a) and in normalized IQ-plane (b).

demonstrate the importance of a sufficient coupling factor, it also shows the response Δf of the two resonators to a mm-wave signal $\Delta P_{optical}$, which is identical in both cases. The dots and squares indicate the measurement points of the readout electronics in the S_{21}- and IQ-plane respectively. Detecting an mm-wave signal, regarding the S_{21}, means a shift in resonance frequency, amplitude and phase (not shown). In the IQ-plane on the other hand, we measure a shift along the resonance circle (corresponding to a phase shift) with a decreased radian until the off resonance circle is reached or the radian is zero. A frequency sweep of a resonance in S_{21} corresponds to a sweep in the IQ-plane as indicated from f_1 to f_3, where f_2 stands for the frequencies on the resonance circle. For an identical signal, the resonance frequencies of both resonators shift by the same value to lower frequencies. The difference between the two resonators is the readout point. For the strongly coupled resonator, the signal is still measurable because the measurement point has not yet reached the 0 dB line in the S_{21}-plane or the off-resonance circle in the IQ-plane. The signal of the weakly coupled resonator is not measurable anymore because the measurement point corresponds to the 0dB line or the off-resonance circle. In that case we speak of a saturated detector because for higher optical loads, we always measure the same value. A weakly coupled resonator is therefore highly sensitive because small signals already lead to a large phase response looking (compare phase responses in

IQ-plane). Its dynamic range on the other hand is limited to a relatively small signal range. A strongly coupled resonator has a large dynamic range but is less sensitive to small signals. Therefore, the coupling factor κ has to be chosen accordingly to the demanded specifications: A compromise between sufficient dynamic range and detector sensitivity.

Another factor that has to be taken into account and also a reason why we need a sufficient coupling strength is the sky background of around T=50 K at the 30 m telescope. Without detecting any particular signal from an astronomical source, the LEKIDs already see a relatively high black body radiation load due to the atmospheric background, leading to a change in resonance frequency and intrinsic quality factor Q_0. The photons of this background radiation excite quasi-particles in the superconducting microresonator increasing the kinetic inductance and the surface losses R_s.

Typical coupling quality factors for the NIKA specifications are between $Q_C = 30000$ and $Q_C = 50000$, which guarantees sufficient high Q_L with sufficient deep resonances. These values have been determined empirically by testing LEKIDs under different mm-wave loadings. The LEKIDs presented here are therefore designed, by simulations and low temperature measurements, with coupling quality factors in that range. Fig. 6.5 shows simulations of an LEKID coupled with different κ to a transmission line. To change the coupling strength, m is varied from 6 to 14 μm whereas d was kept unchanged in this case (see Fig. 6.2). The simulation software *Sonnet* was used for all simulations in this chapter that allows to simulate the superconducting properties of the thin films. They can be defined by the surface resistance and the kinetic inductance as described in 4.2. We see, that not only the shape of the resonance curves changes for different κ, but also the resonance frequency shifts due to the mutual inductance M that makes part of the LC-circuit of the resonator. The unloaded quality factor Q_0 is the same for the three couplings as expected because it only depends on the properties of the resonator itself and is independent of the external network for single resonators. We will see later that Q_0 can change due to cross coupling effects between neighboring resonators but has no influence here for the design of the coupling strength. Assuming a frequency spacing between neighboring resonators of around 2 MHz, the coupling should not exceed a κ of 7 in order to avoid overlaps. We see from the simulation, that the coupling should neither be smaller than $\kappa = 2$ to provide a sufficient dynamic range

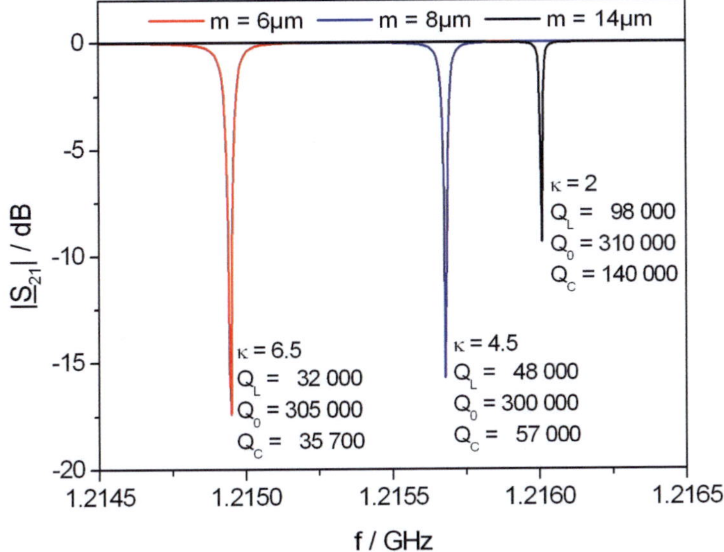

Fig. 6.5.: Simulations of a LEKID coupled to a transmission line for three different coupling factors determined by varying the ground plane width *m*.

regarding the optical background load. With $\kappa < 7$, the bandwidth of the resonance at the carrier amplitude (at 0 dB) is less than 500 kHz and is therefore well adapted to our application.

These simulations have been verified by low temperature measurements using the six pixel array shown above with a 60 nm thick Nb film. For the two results we show here, the distance between resonator and ground plane is kept unchanged to $d = 10\ \mu m$ whereas the ground plane width is $m = 10\ \mu m$ and $m = 20\ \mu m$ (see Fig. 6.2). In Fig. 6.6 the measurement of these two resonators with two different couplings is shown. For visibility reasons, here a zoom on one of the six resonators of each coupling strength is shown. As seen from the simulations, the resonance frequency shifts to lower frequencies for a higher κ due to the mutual coupling inductance M. Q_0 and therefore Q_L is lower than expected from the simulations caused by higher losses in the Nb film than assumed in the simulations. This also leads to differences in the coupling quality factors Q_C for the same coupling factor κ. Cooling these too samples down to $T << 2\ K$, would increase Q_0 and so Q_L,

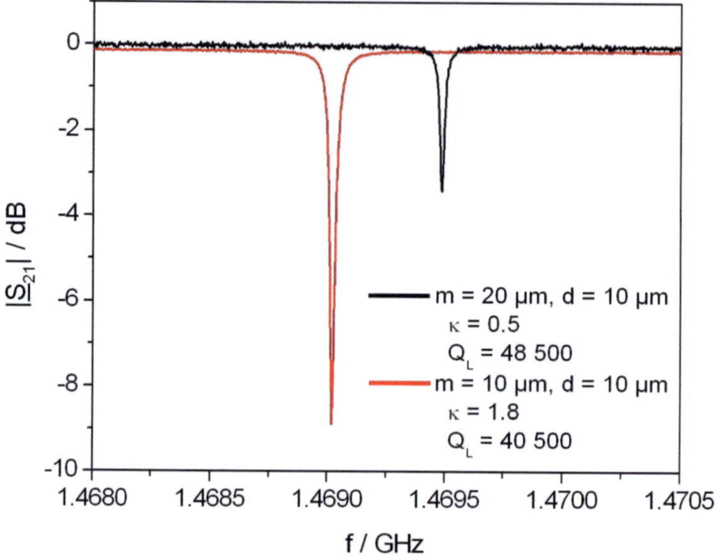

Fig. 6.6.: Measurement of two Nb-LEKIDs with two different couplings.

leading to values of Q_C that are comparable to the simulations. Remember that Q_0 has to be much higher than Q_C to achieve $Q_L \approx Q_C$. Here, taking the Nb sample measured at $T = 2\,K$, this is not the case and thus it is difficult to compare the coupling quality factor Q_C of simulation and measurement. But the results show that the *Sonnet* simulation tool is useful for the design of the coupling strength for these LEKIDs used at sub Kelvin temperatures.

6.2. Frequency tuning

Another important factor to optimize is the frequency tuning of a large number of resonators that are equally spaced in a limited bandwidth, which is important for the readout of these detectors. Resonance frequencies that are too close (overlapping resonance dips) cause errors during the data processing of the readout process and lead to imaging errors of the observed sources. If the frequency spacing is too large, less resonators can be packed in a limited bandwidth. The resonance frequency of a LEKID can be changed by varying the finger length and finger number of the interdigital capacitor. As we have seen above, the resonators should not be packed closer in frequency than 1.5-2 MHz in order to avoid overlapping resonances. The available bandwidth in our case is limited by the AD-converter and is around 230 MHz. With a frequency spacing of 2 MHz, we can therefore theoretically readout around 115 pixels. We show here the simulation and measurement results of two six pixel arrays with different frequency spacing. The geometrical dimensions of the capacitor and the corresponding meander line are shown in Fig. 4.8. The shift in frequency per unit length for this capacitor has been simulated dependent on Δx. It has been determined by simulations that the resonance frequency shifts around 1 MHz for $\Delta x = 30\ \mu m$. Starting with the first finger, the length is being shortened according to the spacing. Reaching the end of one finger, we continue by shortening the second one and so on. Two samples have been designed and measured with different frequency spacings, where Δx has been varied linearly as shown in table 6.1. The 6 resonances of *sample*1 are distributed in around 210 MHz, whereas

$\Delta x/\mu m$	$\Delta x1$	$\Delta x2$	$\Delta x3$	$\Delta x4$	$\Delta x5$	$\Delta x6$
*sample*1	0	300	3000	3300	6000	6300
Δf/MHz	0	10	100	110	200	210
*sample*2	0	600	6000	6600	12000	12600
Δf/MHz	0	20	200	220	400	420

Table 6.1.: Frequency tuning of two six pixel arrays (*sample*1 and *sample*2) with different frequency spacing.

for *sample*2 a larger bandwidth is used (420 MHz). This has been done to verify the linearity of the frequency tuning over a large bandwidth. The geometry of the simulated 6-pixel array is shown in Fig. 6.7, which is a relatively large structure to simulate, especially with a smallest mesh cell of 2 μm, corresponding to half of the meander line width. The frequency resolution could not be set too high in order to achieve a reasonable simulation time. This can lead to different shapes of resonance curves comparing measurement and simulation. Here we are determining the frequency tuning or in other words the resonance frequencies, where the shape of the curve is less important and can therefore use this simulation model. The pixel surrounding ground plane is used as electromagnetic shielding between neighboring resonators and is discussed in more detail later on. As for the determination of the coupling, the samples have been fabricated with a 60 nm thick Nb film and mea-

Fig. 6.7.: Sonnet simulation model of a 6-pixel LEKID array. Frequency tuning from lowest (1) to highest resonance frequency (6).

sured at a temperature of 2 K. The measurement and simulation results of *sample*1 are compared in Fig. 6.8. The simulated resonances are less deep than the measured ones, which is due to the mentioned reduced frequency resolution used in the simulation. Regarding from the frequency tuning point of view, the simulations agree very well with the measurements. The highest deviation between measurement and simulation is less than 10 MHz for the two resonances with the highest resonances frequencies. The good agreement between both allows to simulate and to determine the frequency tuning of future LEKID geometries without the need of fabricating and measuring the samples, which is less time consuming. We also see from the plots that the tuning is not perfectly linear but the spacing gets larger for decreasing capacity. The designed spacing between resonator 1 and 3 of 100 MHz (see 6.7 and table 6.1) turns out to be in the range of 120 MHz. This shift is even larger between the resonators 3 and 5. With the same designed spacing, we measure and simulate a frequency shift of around 160 MHz. The number of fingers of the interdigital capacitor is less than six full lines for the highest frequency of resonator 6. This non-linear effect becomes even more obvious by regarding the results of *sample*2, where the bandwidth is much larger.

In Fig. 6.9 the simulation and measurement results of *sample*2 are shown and compared. At the lowest two resonances, the measured and simulated spacing corre-

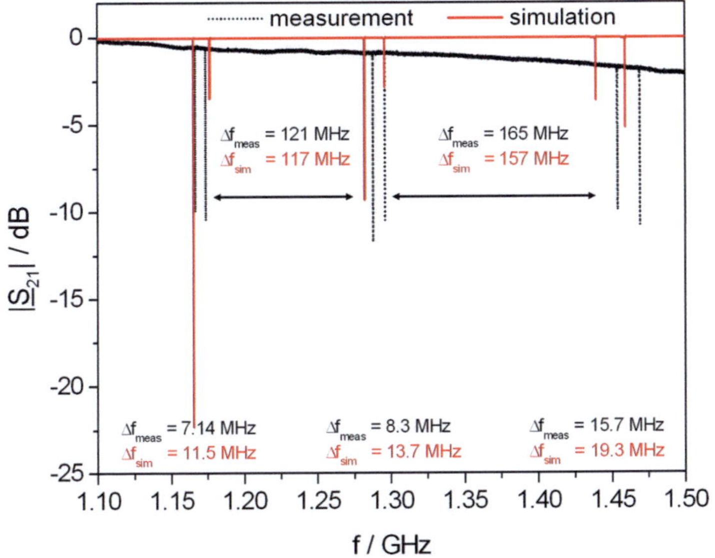

Fig. 6.8.: Measurement (dotted black line) of the six pixels Nb-array *sample*1 at 2 K compared to simulation (solid red line). The plot shows the simulated (red) and the measured (black) frequency spacings Δf_{sim} and Δf_{meas} between the six resonators.

sponds to the designed one. But going to higher frequencies we see that the spacing increases continuously. Between the resonances 3 and 5 with a designed spacing of 200 MHz, we measured around 700 MHz, which is more than a factor three of deviation. For the sixth resonance, the interdigital capacitor is reduced to less than three full finger lines from the original nine fingers ($\Delta x = 0$). Comparing the results of both samples we see that the non-linearity of the frequency tuning depends on the number of the fingers.

To define a minimum number of fingers, where the frequency tuning is relatively linear, the resonance frequencies of both samples have been put in one graph (see Fig. 6.10) to show the behavior of the frequency tuning dependent of Δx for this particular geometry of the interdigital capacitor. We see that the resonance frequency increases with a square root behavior for a linear decreasing Δx. For a finger number between six and nine, the tuning is still relatively linear but shows already an non-linear behavior. For less than six fingers, the non-linearity increases rapidly.

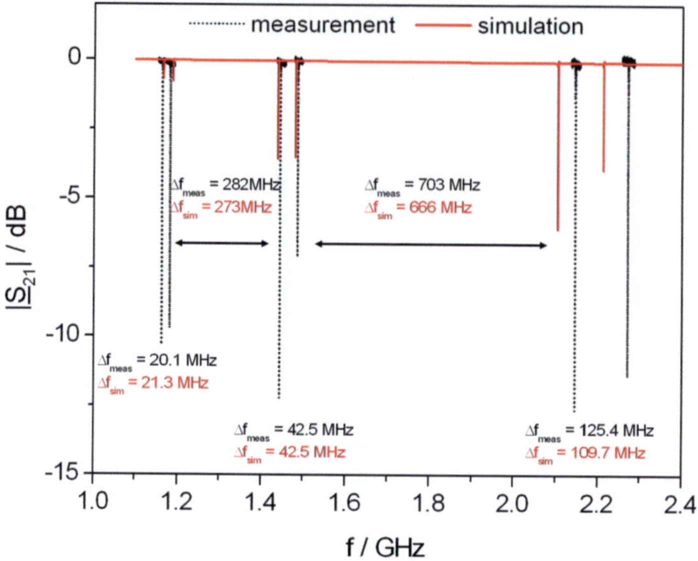

Fig. 6.9.: Measurement (dotted black line) of the six pixels Nb-array *sample*2 at 2 K compared to simulation (solid red line). The plot shows the simulated (red) and the measured (black) frequency spacings Δf_{sim} and Δf_{meas} between the six resonators.

In that range, a constant frequency tuning with equally spaced resonances is difficult to achieve due to a high variation in resonance frequency for small changes in finger length. For the LEKID arrays presented here, we use therefore at least four fix fingers and change the resonance frequency, depending on the geometry and the material of the capacitor, over the length of the additional fingers. This of course has to be taken into account by designing large arrays with hundreds of pixels over a large frequency bandwidth. A linear shortening of finger length would lead to a much wider needed bandwidth and a large part of it could not be used efficiently. In smaller bandwidths of around 200MHz, which corresponds to the available NIKA bandwidth, the frequency tuning can be considered as linear as a good approximation. We do not give a general equation for the fit shown in Fig. 6.10. The frequency tuning depends strongly on the geometry of the interdigital capacitor (finger width, finger spacing and finger length) and is therefore different for each capacitor structure. Here, we want to show the non-linear behavior of the frequency tuning by

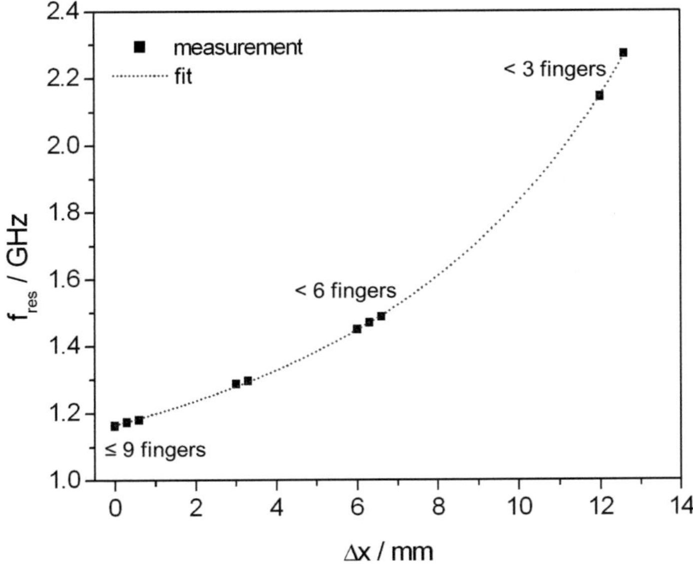

Fig. 6.10.: Measured resonance frequencies of *sample*1 and 2 dependent of Δx (black squares). A fit is plotted over the measured resonance frequencies to show the square root behavior of the resonance frequency tuning over the linear change in finger length.

changing the length of the capacitor fingers that becomes more important for arrays with much more than 115 pixels.

6.3. Electromagnetic pixel cross coupling

After characterizing the LEKID resonance circuit by its parameters as resonance frequency, quality factor and coupling, it is also important to optimize the geometrical distribution and packaging of the resonators on an array. The filling factor has to be sufficiently high to reduce dead area on the wafer but on the other hand, the LEKIDs cannot be packed arbitrarily dense due to a relatively strong electromagnetic cross coupling between the pixels. The resonators possess a certain radiation loss and if the coupling between two pixels is too high, one resonator influences the resonance frequency and the quality factor of a second one leading to a non-homogeneous frequency distribution. To minimize these effects, a ground plane fra-

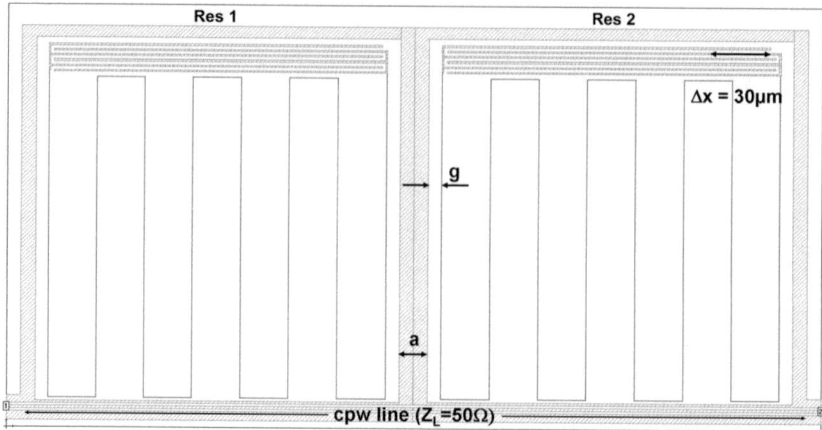

Fig. 6.11.: *Sonnet* simulation model for two neighboring LEKIDs surrounded by a ground plane frame and coupled to a coplanar transmission line. The difference in resonance frequency is kept relatively small with around 1.5 MHz corresponding to $\Delta x = 30\ \mu m$. The varied parameters here are the spacing g between resonator and ground plane and the width of the ground plane a. The six interdigital capacitor fingers of each resonator are 10 μm wide, 20 μm spaced and 1900 μm long.

me is put around each pixel as an electrical shielding. One problem with this ground plane is the influence on the designed resonance frequency. The ground plane acts as an anode of a capacitor around the resonator and leads therefore to an additional capacitance between resonator and ground plane. But since this behavior is the same for each pixel, it leads in fact to a parallel shift in resonance frequency for all resonators on the array. To investigate effects such as these frequency shifts and electromagnetic cross coupling, simulations of two neighboring LEKIDs, as shown in Fig. 6.11, have been done. The parameters with the most influence on these effects are given by the distance g between the resonator and the ground plane with line width a. The shown LEKID geometry is adapted to aluminum as superconducting material. The coupling to the transmission line is designed to be $Q_C = 30000$ and the interdigital capacitor consists of six fingers, 10 μm wide and 20 μm spaced. Measurements have shown that the noise of the LEKIDs can be decreased by a wider spacing of the fingers due to a lower electrical field strength reducing the dielectric losses in the substrate [46]. Due to a higher kinetic inductance of aluminum, the finger number can be reduced to only six to determine the resonance frequency in the same frequency band as the Nb resonators. The two resonators are

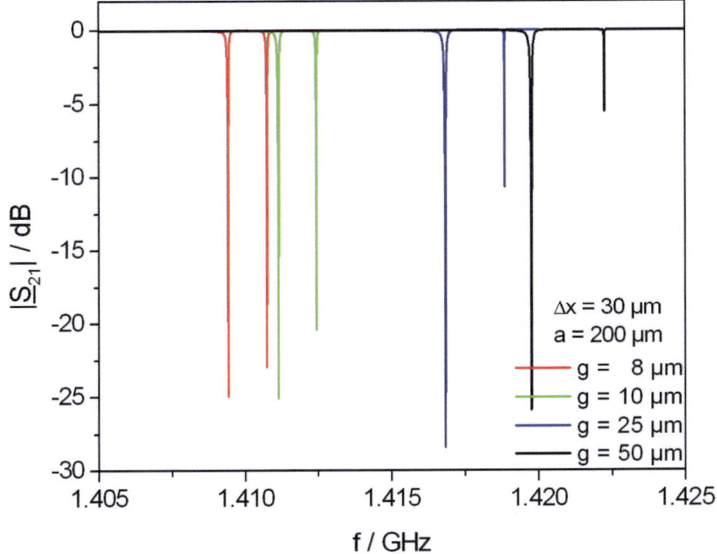

Fig. 6.12.: Influence of the ground plane distance g to the resonance frequencies of both resonators. The frequency spacing is designed to 1.5 MHz ($\Delta x = 30\ \mu m$) and the ground plane width is $a = 200\ \mu m$.

spaced by $\Delta f \approx 1.5\ MHz$ corresponding to $\Delta x = 30\ \mu m$. This frequency spacing is relatively close and is chosen to investigate the cross coupling effects, which are higher when the two resonance frequencies are closer.

The first simulation has been done to investigate the influence of the ground plane distance to the resonator, where the ground plane width of $a = 200\ \mu m$ has been kept unchanged and the spacing between resonator and ground plane ($g = 8, 10, 25, 50\ \mu m$) is varied. The results of these simulations are shown in Fig. 6.12. We see a large shift to lower resonance frequencies for decreasing g. This confirms the assumption of a capacitance between resonator and the ground plane, which increases when the resonator gets closer to the ground plane.

We also notice that the shape of the resonances deviates from each other for lower g and the frequency spacing increases to higher values than the designed 2 MHz. It seems, that a larger spacing ($g = 50\ \mu m$) enhances the electromagnetic coupling between the resonators. The two asymmetrical resonance curves are typical for such cross coupling effects between two resonators. It shows that the second resonator is excited over the radiation losses of the first one depending on the distance and the ground plane between the two. This adds another coupling to the second resonator than it would be from the transmission line only and changes the original designed resonance frequency due to a variation in total impedance. Part of the energy stored in one resonator couples therefore to the second one and changes the quality factor Q_0 as seen in Fig. 6.12. If g is small, the electrical field between resonator and ground plane increases, leading to a higher capacitance and shifting f_{res} to lower frequencies. But on the other hand, the short distance reduces the electrical field density to a smaller volume in the air above the resonators and below in the substrate. This means, that the coupling of electrical fields between the resonators is relatively small. This effect is much stronger in the substrate below than in the air above the resonator due to the higher dielectric constant. The principle of this effect is shown as a schematic in Fig. 6.13, where we see that the cross coupling can be

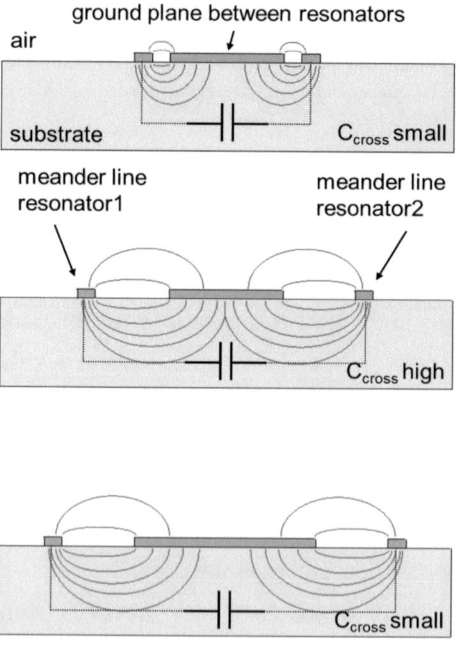

Fig. 6.13.: Schematic to demonstrate the influence of the resonator geometry to the cross coupling between them defined by g and a.

reduced by changing the two geometrical parameters a and g. The schematic shows the cross section of the ground plane and the two closest meander lines of each resonator. The cross coupling effects can hereby be reduces by increasing the ground plane width a (case c)) and decreasing g (case a)). The most electromagnetic coupling between two resonators is found in case b), where a is narrow and g relatively wide.

In theory we can now design the resonators as close as possible to a large ground plane to minimize these effects. But in reality we have to think about the consequences caused by this geometry. If the ground plane gets larger and larger, we lose a lot of space on the LEKID array that could be filled with detectors, which reduces the optical efficiency of the array and makes the imaging process more complicated. Therefore a reasonable pixel filling factor on the array has to be taken into account by designing them. The second concern is the distance between resonator and ground plane. If it gets smaller and smaller, the capacitance between resonator and ground plane increases and exceeds the capacity of the interdigital capacitor. In that case the determination of the resonance frequency is dominated by this capacitance and a regular frequency tuning becomes difficult. Another reason why this capacitance has to be kept smaller than that of the interdigital capacitor is the quality factor and the current distribution in the meander section. The loaded quality factor Q_L would decrease due to a power flow to the ground plane or in other words less power can be stored in the resonator decreasing Q_0. If less power is stored in the resonator, the current density along the meander line is smaller and the detector less sensitive. For the design of large detector arrays with small pixel crosstalk, these mentioned parameters have to be well adapted and optimized.

To simulate the influence of both a and g to the cross coupling we can use the same model as before, but this time the two resonators have the same resonance frequency with $\Delta x = 0\ \mu m$. In an ideal model without any cross coupling, we should see only one resonance with lower Q_0 than that of a single resonator. So half of the power is stored in each resonator at the same frequency. If there is any cross coupling, two resonances will appear due to the additional capacitance C_{cross}, changing the resonance frequencies. The higher the cross coupling, the higher the frequency spacing Δf_{cross} between the two resonances and the higher the difference in depth of the resonance dips A_{cross} due to a lower Q_0. In Fig. 6.14, the results of such simulations are demonstrated. a was kept unchanged at 200 μm and g is varied from 8 to 50

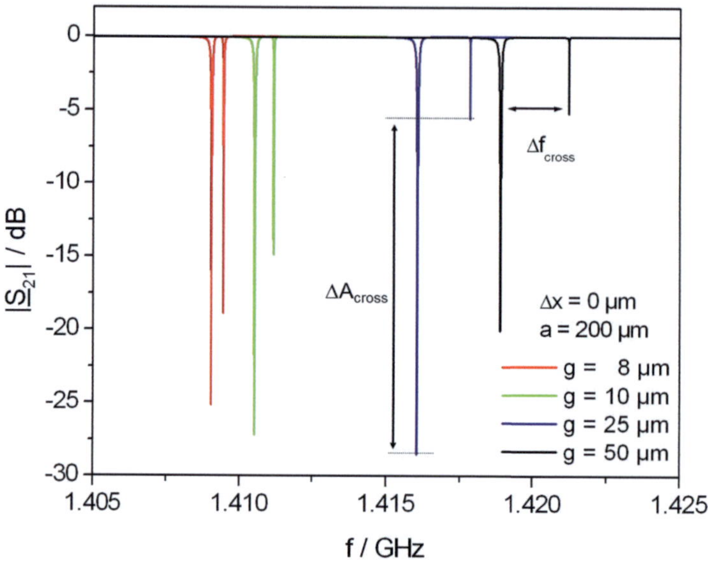

Fig. 6.14.: Simulation of two neighboring resonators with the same resonance frequencies ($\Delta x = 0\ \mu m$) to demonstrate the influence of the distance g between resonator and ground plane to the electromagnetic pixel cross coupling.

μm. For $g = 25$ and $g = 50\ \mu m$, the cross coupling is much higher than for smaller g, which can be seen from the relatively large spacing Δf_{cross} compared to that with $g = 8\ \mu m$. Also the shape of the resonance curves is, as expected, very different (see ΔA_{cross}). This confirms the assumption, that power from one resonator couples into the second one decreasing Q_0. The higher Δf_{cross} and ΔA_{cross}, the higher C_{cross}. Even for a very small g there are still two resonances, but with a smaller Δf_{cross} and ΔA_{cross}. For such large field detector arrays that we want to build, there will always be a certain cross coupling between the pixels, but this gives us the possibility to reduce it as much as possible. Simulations have also shown, that g should not be smaller than $10\mu m$ to guarantee high quality factors and high current densities in the LEKID meander lines.

The other geometrical factor that is directly linked to C_{cross} is a, the width of the ground plane between the resonators. g is therefore kept unchanged for the simulations and a is varied this time. The resonance frequencies of both resonators are

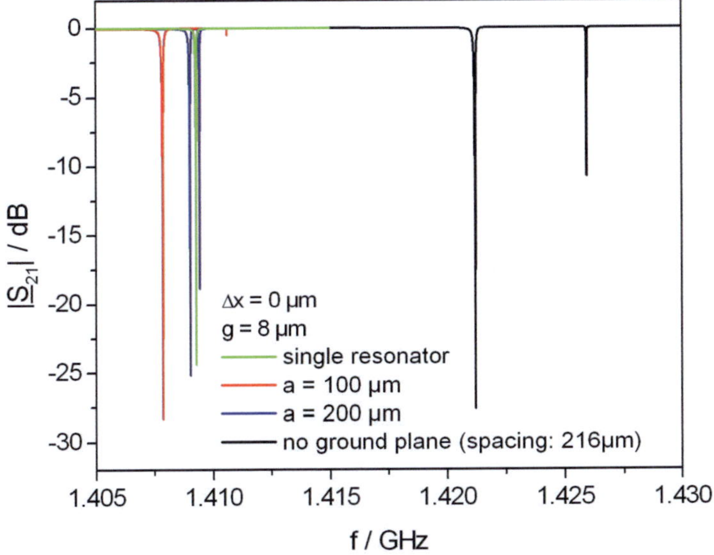

Fig. 6.15.: Simulation of two identical neighboring resonators with $\Delta x = 0$ to demonstrate the cross coupling effect dependent of the resonator environment. For comparison reasons, the simulation of a single resonator of the same design is added (green line).

identical ($\Delta x = 0\ \mu m$) also for these simulations. The plots in Fig. 6.15 show simulations with two different ground plane widths, $a = 100\ \mu m$ and $a = 200\ \mu m$. Two other simulations have been added to show the shielding effect of the present ground plane. One without a ground plane in between the resonators and a second one of a single resonator with the same geometry to compare the *real* resonance frequency without any cross coupling effect. The simulations show, that most of the cross coupling can be reduced by adding a ground plane between the resonators. Δf_{cross} decreases from around 5 MHz to around 400 kHz due to this shielding. As expected increases the cross coupling with decreasing a because the resonators are closer, which increases C_{cross}. For comparison reasons and to demonstrate the influence of the ground plane between two resonators with different resonance frequencies, the same simulations have been done with $\Delta x = 30\ \mu m \rightarrow \Delta f \approx 1.5\ MHz$. The result is shown in Fig. 6.16, where we see that the cross coupling between the resonators

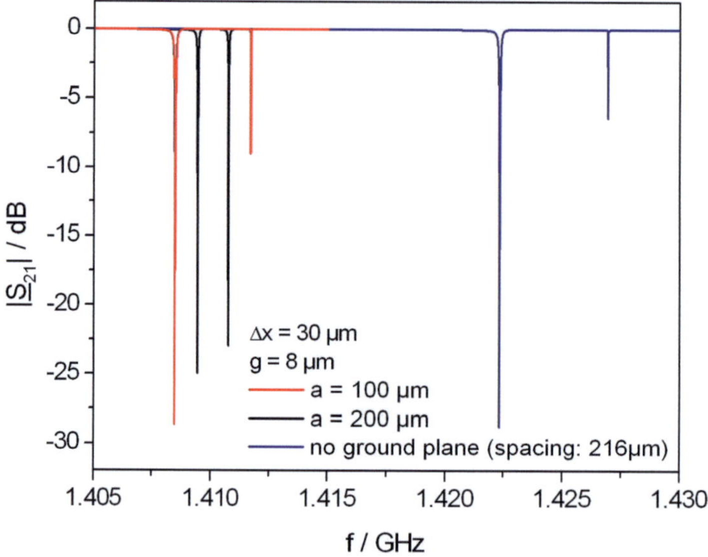

Fig. 6.16.: Simulation of two neighboring resonators with different resonance frequencies $\Delta x = 30\ \mu m \rightarrow \Delta f \approx 1.5\ MHz$. Here the ground plane width a is varied and $g = 8\ \mu m$ is fix.

can be much improved due to the presence of a ground plane. It also confirms, that the pixel crosstalk reduces for larger ground planes as seen above. In order to keep the pixel size at a minimum, a cannot be increased arbitrarily. Otherwise the filling factor of the array becomes too small and the efficiency of the pixel array degrades. The only way to achieve $C_{cross} \rightarrow 0$ would be a resonator with $Q_L \rightarrow \infty$. The development of large arrays is therefore a tradeoff between array filling factor and acceptable electrical cross talk between the pixels.

6.3.1. Modeling the electromagnetic cross coupling

The simulations above have shown, that we can consider the cross coupling as a parasitic capacitance between two resonators, which gives us the possibility to model the electromagnetic cross coupling with the equivalent circuit schematic shown in Fig. 6.17, where the cross coupling is represented by such a capacitance C_{cross}. In this model we neglect the ohmic losses and consider a pure reactive LC-circuit.

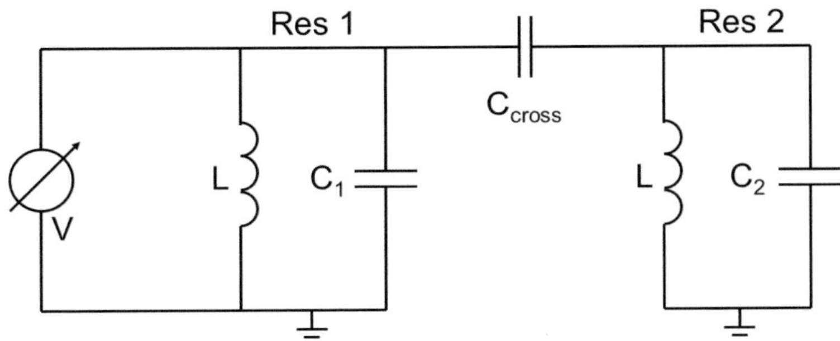

Fig. 6.17.: Equivalent circuit of two resonators with parasitic cross coupling capacitance C_{cross}. If the resonant circuit of $Res1$ is on resonance, it can be considered as voltage source due to the high potential difference over the inductance or the capacitor.

If one resonator ($Res1$) reaches its resonance frequency, it can be considered as a voltage source for the second one ($Res2$) due to a high electrical potential over either the inductance or the capacitance. If $C_{cross} \neq 0$, the voltage is applied on the impedance of $Res2$ plus C_{cross}. This way, it is possible to calculate the resonance frequency of both resonators dependent of C_{cross}. Thus we have to calculate the impedance of this circuit, which is given by

$$Z = \frac{j\omega L}{1-\omega^2 LC_1} + \frac{1}{j\omega C_{cross}} + \frac{j\omega L}{1-\omega^2 LC_2} \tag{6.3}$$

The inductance L is the same for both resonators since the resonance frequency is tuned by the capacitor and the meander structure is unchanged. We assume here perfect conductors to reduce the calculation and Z becomes therefore purely imaginary. As explained before the imaginary parts of a resonator (L and C) cancel each other out, when the resonator is on resonance. We can therefore derive the two resonance frequencies, by calculating the zeros of this imaginary impedance Z leading to the following equation:

$$\begin{aligned} 0 &= \omega^4 - \omega^2 \frac{C_1 + C_2 + 2C_{cross}}{L(C_1C_2 + C_1C_{cross} + C_2C_{cross})} \\ &\quad + \frac{1}{L(C_1C_2 + C_1C_{cross} + C_2C_{cross})} \end{aligned} \tag{6.4}$$

By substituting $\omega^2 = z$, the quadratic equation can be solved easily and the two resonance frequencies can be calculated with $f_{res} = \omega/2\pi$.
To calculate them, we must first determine the value of C_{cross}. Assuming, that most of the cross coupling is caused by the two closest meander lines of the two LEKIDs, a model as shown in Fig. 6.18 has been simulated to determine C_{cross}. It consists of a ground plane plane that surrounds the two pixels as seen above and the two closest meander lines of $Res1$ and $Res2$ simulating the capacitance C_{cross} between two pixels, where the parameters to vary are g and a. The length of the meander line section corresponds to the real LEKID side length of 1850 μm. From the simulated scattering parameters using the model below, we can calculate C_{cross} by considering the circuit as a two port system with serial impedance. We calculate it the same way, using a ABCD matrix, as for the determination of L_{line} and C for the LEKID resonant circuit (see 4.2.1). Several simulations for different values of a and g have been done with a fixed substrate (silicon) thickness of $300 \mu m$.
Fig. 6.19 shows the results of these simulations that can be used to calculate the resonance frequencies dependent of the cross coupling for the circuit in Fig. 6.17.

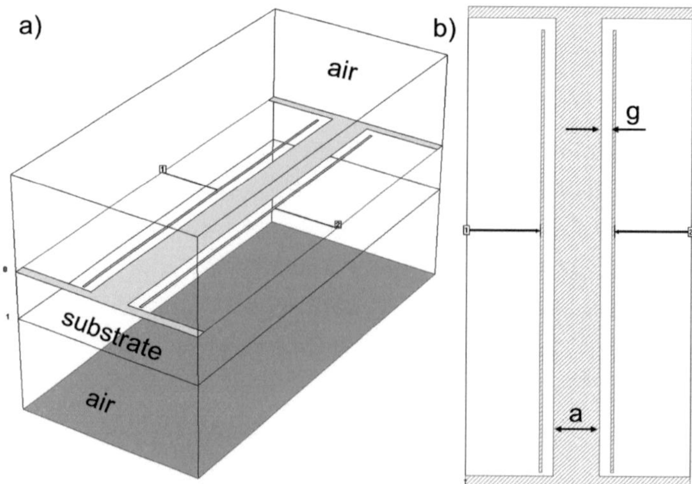

Fig. 6.18.: Sonnet simulation model to determine C_{cross}. a) 3D-model with dielectric layer configuration; b) 2D-model with geometrical parameters a and g. The length of the meander line is 1850 μm.

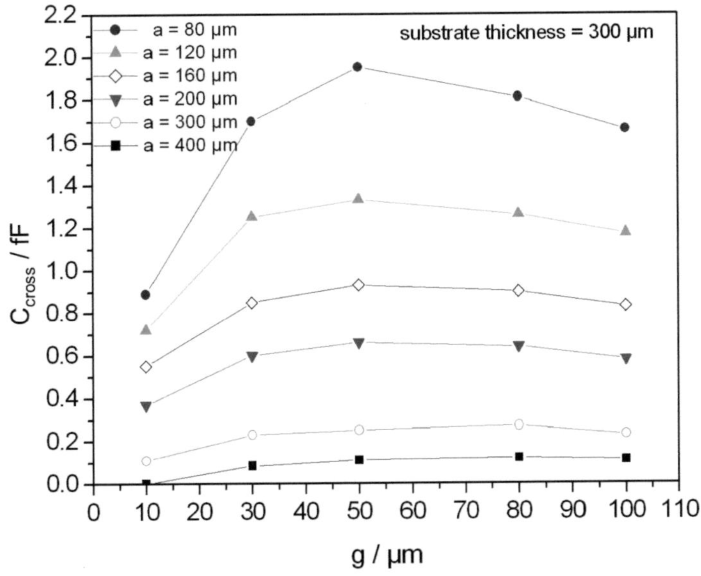

Fig. 6.19.: Calculated parasitic capacitance C_{cross} for different values of a and g. The silicon substrate thickness is 300 μm. The values have been calculated from the simulated scattering parameters.

These results confirm the simulations presented above. The capacity decreases for a larger ground plane and smaller distance of resonator to ground plane. Regarding the dependence of g, we see that C_{cross} increases until its maximum at around $g = 50\ \mu m$. For higher values, it starts to decrease due to the increasing distance that lowers now the strength of the electrical field and therefore the cross coupling effect. Comparing C_{cross} for $a = 80\ \mu m$ and $a = 200\ \mu m$ at $g = 50\ \mu m$, we see that the value is reduced by more than a factor of three.

In Fig. 6.20 the two resonance frequencies are plotted over the simulated values of C_{cross} for a given LEKID structure. The inductance L_{line} and the capacitances C_1 and C_2 have been determined as seen before. For $Res1$ we have chosen a $L_{line} = 26\ nH$ and $C_1 = 0.36\ pF$ and for the second one $C_2 = 0.361\ pF$, which corresponds to $\Delta f \approx 2\ MHz$. For $C_{cross} = 0$, Δf is at its minimum, coincident with the designed frequency spacing. Increasing C_{cross} spreads the resonances further away from each other, corresponding to the simulation results shown above. We see from the si-

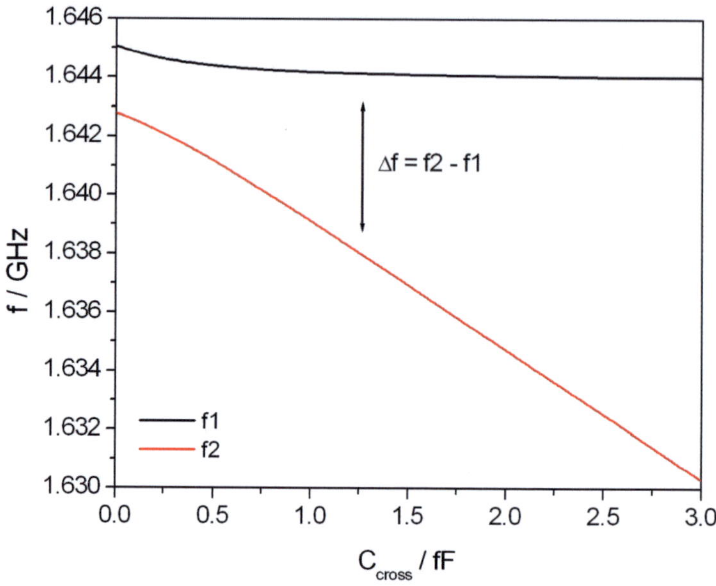

Fig. 6.20.: Calculation of the cross coupling frequency shift Δf dependent of C_{cross}. The curves have been calculated using equation 6.4 and the simulated values for C_{cross} (see Fig. 6.19).

mulations and calculations that the cross coupling has a big influence on the two resonance frequencies and it is therefore of big interest to minimize it, in order to design large pixel arrays. The results from the calculations using the simple model of two resonant circuits connected by an additional capacity agrees very well with the simulations done before including the entire LEKID geometry, which is confirmation of the assumed capacity dominated electromagnetic cross coupling effect. Here only two resonators have been considered but the model in Fig. 6.17 can be extended to see the cross coupling to a third or fourth resonator. But this gets very complicated for more than two resonators and is beyond the scope of this chapter.

Another series of simulations have been done to see the influence of the substrate thickness on C_{cross}. The dielectric constant of the silicon increases C_{cross} and has not been taken into account yet as a possibility to reduce the cross coupling. To reduce the capacitance in the substrate, one option is to decrease the substrate thickness. Using the same model from Fig. 6.18, C_{cross} has been calculated but this time for

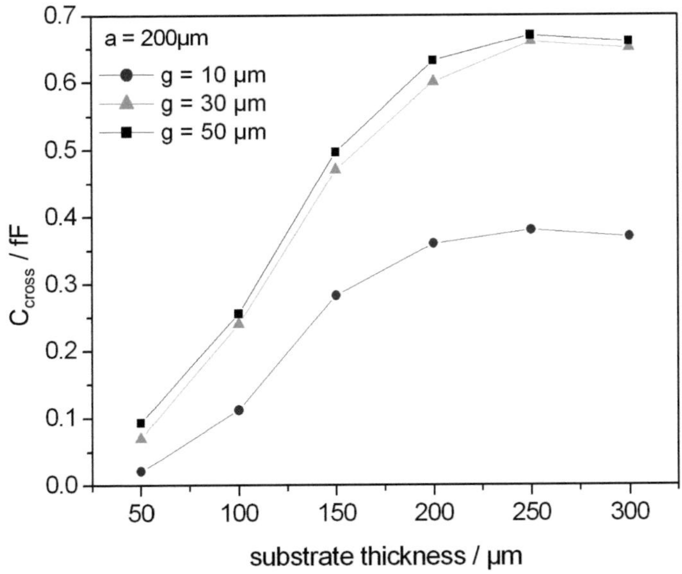

Fig. 6.21.: Calculated parasitic capacitance C_{cross} for different substrate thicknesses.

varying substrate thickness. The results shown in Fig. 6.21 have been simulated for three different values for g and a fix value for $a = 200\ \mu m$. The substrate thickness is varied from 50 to 300 μm. As expected, the capacitance C_{cross} decreases for decreasing substrate thickness due less dielectric material volume and reduces this way the cross coupling. This also confirms the fact that the cross coupling is dominated by the radiation losses in the substrate caused by the higher dielectric constant. The substrate thickness plays a major role for the optical coupling of LEKIDs and can therefore not been varied arbitrarily. This is discussed in more detail in the following chapter. Substrates that are thinner than 100 μm are in general difficult to handle regarding the wafer processing in the clean room and the assembly of the array into the sample holder. There are several other way to decrease the cross coupling capacity C_{cross} by modifying the substrate. One possibility is to etch in between the resonators into the silicon substrate. This way, the inside of the trench could be metal-coated via a sputtering process to improve the electromagnetic shielding between the resonators. This has not been simulated or realized yet and will be subject of future work to optimize the cross coupling problem.

6.4. Geometrical LEKID distribution on multi pixel arrays

A second consequence of pixel cross coupling that has not been mentioned yet arises when photons are absorbed in the resonator. This effect is of course linked to the cross coupling above, but has different impact. If a LEKID absorbs photons, it changes its resonance frequency, which is the detecting principle of these detectors. If the coupling to the neighboring resonator is strong enough, it affects also the resonance frequency of this resonator even if he does not absorb any photons, leading to imaging errors.

One possibility to improve this pixel cross coupling effect can be achieved by the geometrical pixel distribution on the array. The influence of C_{cross} increases if the neighboring resonance frequencies are very close, which can be derived from the model of two neighboring LC resonators connected by C_{cross} shown above. We avoid this by distributing the resonators in a way that neighboring resonators on the array are not neighbors in resonance frequency. We arrange therefore the pixels

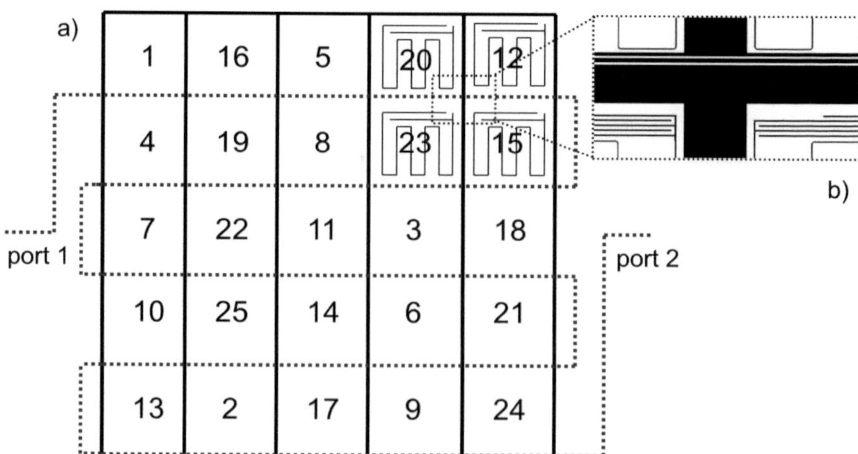

Fig. 6.22.: a) Geometrical resonator arrangement of a 25 pixel LEKID array to a transmission line (red). The numbers correspond to the order of resonance frequencies starting with the lowest at number 1. b) Zoom to the coupling area of two resonators.

with a maximum possible frequency spacing to each other all over the array. One example of a 25 pixel array is shown in Fig. 6.22a). Here the minimum resonance frequency spacing between neighboring resonators is at least three frequency steps. This is only one example of how the pixels can be arranged. Measurements have shown that this kind of cross talk can be much improved by the proposed resonator arrangement due to its dependence on the frequency spacing to neighboring resonators.

The red line in the schematic represents the transmission line to which the resonators are coupled to. The resonators are coupled from only one side to the line. Measurements and simulations have shown that cross coupling effects across the transmission line can be reduced this way. Fig. 6.22b) shows a zoom to the coupling area of two pixels across the transmission line. The ground plane width above the center line, where the pixel is actually coupled to, is narrower and defines the coupling strength. The larger ground plane below the center line avoids the excitation of a second resonator over the interdigital capacitor and at the same time reduces the cross coupling across the transmission line. The ground plane on the lower side has to be chosen wide enough so that the current distribution is approximately zero to reduce the mutual inductance between the transmission line and the resonator below (see current distribution on a cpw-line in 4.2). All samples of LEKID arrays that are measured and discussed later on, are designed and fabricated with this kind of transmission line guidance and pixel arrangement.

7. Optimization and measurement of the mm-wave coupling

After determining and optimizing the electrical properties of LEKIDs, such as resonator coupling, resonance frequency and cross-talk issues, we have to investigate the mm-wave coupling. Instead of mm-wave coupling, it is common to use the word *optical coupling* although we are not considering the optical spectrum. As already mentioned, one of the advantages of using LEKIDs is the combination of resonator and direct absorption area in the same structure making the use of antennas or lenses unnecessary. It is therefore important to have a high mm-wave absorption in the LEKID structure, which is directly linked to the sensitivity of the detector. The more power is coupled to the device and absorbed, the higher the response of the detector or, in other words, the higher the signal to noise ratio. The coupling efficiency depends in the LEKID case on the geometry of the meander line and the normal resistivity of the metallization. The principle of the coupling or the absorption of a mm-wave signal in the NIKA LEKIDs case is shown as a schematic in Fig. 7.1.

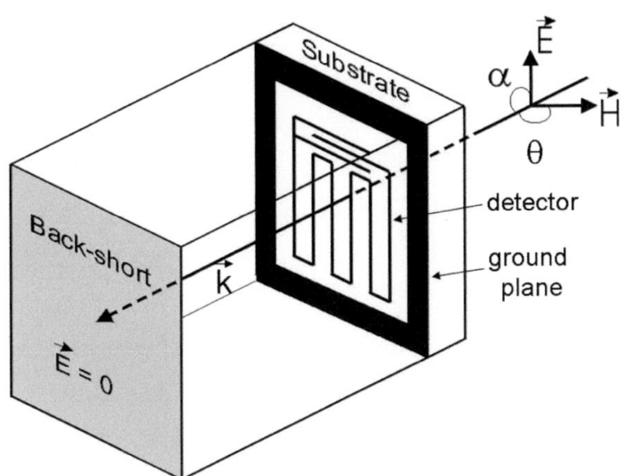

Fig. 7.1.: Principle of coupling a mm-wave signal to a LEKID.

The LEKID is illuminated through the substrate, which means, that the electromagnetic wave passes the substrate before hitting the detector. On the opposite side of the LEKID, a back-short cavity is situated in a calculated distance to the substrate. The substrate thickness and the back-short are important factors by optimizing the optical coupling. In this chapter, a way to model and to optimize it is demonstrated and verified by simulations and measurements.

7.1. Modeling and optimizing the optical coupling

7.1.1. Resistive sheet in free space

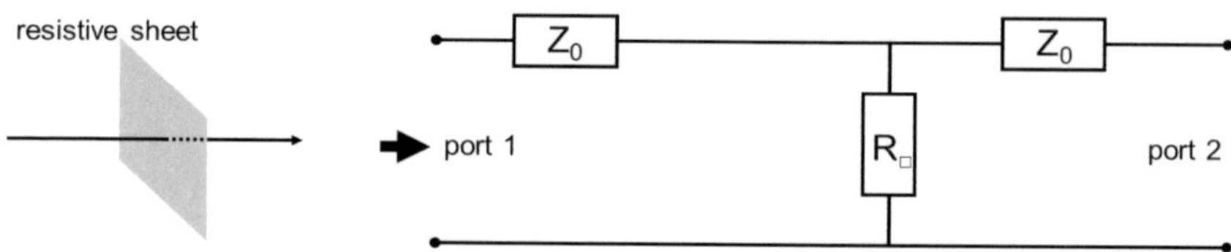

Fig. 7.2.: Schematic and transmission line model of the optical coupling of a resistive metal sheet in free space. The arrow shows the direction of the incoming mm-wave signal passing the metal sheet.

Before starting to calculate the microwave absorption in the relative complex LEKID structure, we will first consider the simplest geometry of a direct absorber: a resistive thin metal sheet in free space. A schematic of this and its equivalent circuit is shown in Fig. 7.2, where Z_0 is the free space impedance and $R_\square$ the sheet resistance of the metal film in Ω/square. In our case, absorption can only be achieved, if the frequency of the photons is higher than the gap frequency of the superconducting aluminum ($h\nu > 2\Delta$)(see Chapter 3), which means that the frequency has to be higher than 90 GHz. In that case, $\sigma_2 \approx 0$ and the complex surface impedance of the aluminum is dominated by $\sigma_1 \approx \sigma_n$. Therefore the sheet resistance is comparable to that in the normal conducting state just above the critical temperature $T_{C,Al} = 1.2\ K$, even if the device is operating at much lower temperatures ($T \approx 100\ mK$). This can be verified by solving the $Mattis-Bardeen\ integrals$ for frequencies with $h\nu > 2\Delta$ [60]. The resistive sheet is modeled in a so-called *transmission line model* as a shunt impedance to ground. The ABCD matrix for such a shunted impedance in a two port system is given by

$$\begin{pmatrix} A = 1 & B = 0 \\ C = 1/R_{\square} & D = 1 \end{pmatrix}$$

From the ABCD matrix we can calculate the S-parameters S_{11} and S_{21} of the two port circuit in Fig. 7.2, which are given by

$$S_{11} = \frac{1}{1+2\frac{R_{\square}}{Z_0}} \quad (7.1) \qquad\qquad S_{21} = \frac{2\frac{R_{\square}}{Z_0}}{1+2\frac{R_{\square}}{Z_0}} \quad (7.2)$$

We can then calculate the absorbed power A in the resistive sheet from the S-parameters. For such a two port system, the absorption is given by

$$A = 1 - (|S_{11}|^2 + |S_{21}|^2) \quad (7.3)$$

This leads in our case, for the resistive sheet in free space, to a absorption A_{fs}.

$$A_{fs} = \frac{4\frac{R_{\square}}{Z_0}}{(1+\frac{2R_{\square}}{Z_0})^2} \leq 0.5 \quad (7.4)$$

The absorbed power in the configuration of Fig 7.2 can reach a maximum of 0.5 for $R_{\square} = Z_0/2$. At this sheet resistance, $R_{\square}$ is perfectly matched to the free space impedance, where one half of the input power is absorbed and the other half transmitted to port 2. Using this configuration is therefore not very efficient. In Fig 7.5 the dependency of the absorption from $R_{\square}$ is demonstrated with a maximum at 189 Ω. We see, that the impedance mismatch is worse for $R_{\square} < Z_0/2$ than it is for $R_{\square} > Z_0/2$.

7.1.2. Resistive sheet in free space with back-short cavity

To achieve absorption in R, that is higher than 50 %, we have to design a circuit, where no power is transmitted to port2 and all the input power from port1 is dissipated in $R_{\square}$. To do so, the impedance at port 2 has to be transformed to a much higher value, close to infinity. This can be realized by adding a back-short cavity behind the resistive sheet with a distance of $\lambda_0/4$, where λ_0 is the free space wavelength of the incoming photons. In Fig. 7.3 the schematic and the equivalent circuit for this

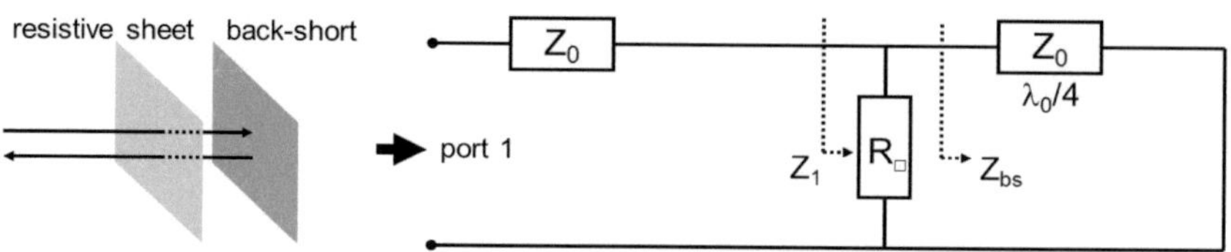

Fig. 7.3.: Schematic and transmission line model of the optical coupling of a resistive metal sheet in free space short ended with a back-short cavity with distance of $\lambda_0/4$. The arrow shows the direction of the incoming and reflected mm-wave signal.

configuration are shown. The back-short is modeled as a short ended transmission line of length l_{bs}. This of course makes the absorption frequency-dependent and has to be taken into account for the calculation of the optical coupling. The input impedance of such a transmission line is given by

$$Z_{bs} = jZ_0 tan(\beta_0 l_{bs}) \tag{7.5}$$

where $\beta_0 = 2\pi/\lambda_0$ with the wavelength in free space λ_0. Remember, that the NIKA LEKIDs are designed primary for the frequency band from 125 to 175 GHz with center frequency at around 150 GHz. In our case, λ_0 is therefore 2 mm, which corresponds to 150 GHz and will be taken for the resistivity dependent calculations. From 7.5 we see, that the input impedance of the short ended transmission line goes to infinity for $l_{bs} = \lambda_0/4$ (500 μm). If we calculate now the input impedance of the back-short combined with the resistive sheet to:

$$Z_1 = \frac{1}{\frac{1}{R_\square} + \frac{1}{Z_{bs}}} \tag{7.6}$$

we can see that, for $l_{bs} = \lambda_0/4$, Z_1 depends only on $R_\square$. The reflection scattering parameter S_{11} for this configuration is given by

$$S_{11} = \frac{Z_1 - Z_0}{Z_1 + Z_0} \tag{7.7}$$

and the absorbed power in the one port circuit can therefore be calculated with

$$A = 1 - |S_{11}|^2 \leq 1 \tag{7.8}$$

If Z_1 depends only on $R_\square$, the absorption also depends only on the sheet resistance. In that case absorption up to 100 % can be achieved if $R_\square = 377\ \Omega$ (see (7.7)). As for the previous circuit, the $R_\square$-dependency of the absorption is shown in Fig. 7.5, here for a frequency of 150 GHz and a back-short length of $l_{bs} = \lambda_0/4$ (500 μm). Comparing the two configurations, we see that the maximum absorption can be doubled, but to achieve that, the sheet resistance has to be doubled as well.

7.1.3. Resistive sheet on substrate with back-short

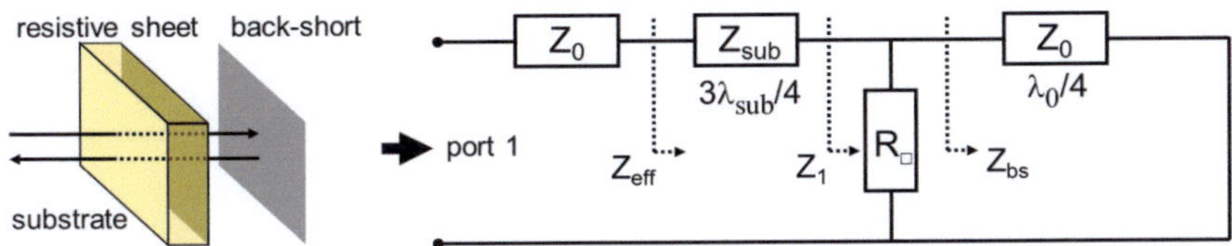

Fig. 7.4.: Schematic and transmission line model of the optical coupling of a resistive metal sheet on a substrate with thickness $3\lambda/4$ short ended with a back-short cavity with distance of $\lambda_0/4$. The arrow shows the direction of the incoming and reflected mm-wave signal.

Sheet resistances of that order are difficult to achieve, even if the film thickness is very small. To keep the maximum absorption of 100 %, the input impedance has to be transformed to a lower value but has to be kept well matched to Z_0. This can be done by using a substrate with a permittivity higher than that of air as a $\lambda/4$-transformer. In Fig. 7.4 the schematic and the equivalent circuit with a substrate in front of the resistive sheet is shown. A transmission line with the length of $(1+2n)\lambda_{sub}/4$ (n=0,1..) can be used to transform a high impedance to a smaller one, where the transformation factor depends on the wave impedance of the transmission line. In our case it is given by

$$Z_{sub} = \sqrt{R_1 R_2} \tag{7.9}$$

where Z_{sub} is the wave impedance of the transmission line (here the silicon substrate), R_1 the resistance that has to be transformed and R_2 the transformed resistance. We can see, that the lower Z_{sub}, the lower R_2. Using a silicon substrate with $\varepsilon_r = 11.9$, the wave impedance can be calculated with

$$Z_{sub} = \frac{Z_0}{\sqrt{\varepsilon_r}} \tag{7.10}$$

and we get for the silicon substrate $Z_{sub} = 109\ \Omega$. From the previous configuration we know that $R_1 = 377\ \Omega$, which corresponds to $R_\square$, where the absorption is at its maximum. With 7.9 we calculate a $R_2 = 31.5\ \Omega$ as new value for $R_\square$. To calculate the absorption the new input impedance Z_{eff} has to be determined. The substrate is, as the back-short, modeled as a transmission line of length l_{sub}. The input impedance Z_{eff} of this configuration can be calculated from the transmission line theory [49] to

$$Z_{eff} = Z_{sub} \frac{Z_1 + jZ_{sub} tan\left(\beta_{sub} l_{sub}\right)}{Z_{sub} + jZ_1 tan\left(\beta_{sub} l_{sub}\right)} \tag{7.11}$$

where $\beta_{sub} = 2\pi/\lambda_{sub}$. λ_{sub} is the modified wavelength due to the dielectric constant of the substrate and is given by $\lambda_0/\sqrt{\varepsilon_r}$. We can now define the new S_{11} as before to

$$S_{11} = \frac{Z_{eff} - Z_0}{Z_{eff} + Z_0} \tag{7.12}$$

The absorption of this circuit, with back-short and substrate, can be calculated as before with (7.8). The resistivity dependence of the absorption of this configuration is also plotted in Fig. 7.5 and can be compared to the others. The frequency is 150 GHz and the substrate thickness is $3\lambda_{sub}/4$ (435 μm). As expected, the maximum absorption is now at $R_\square = 31.5\ \Omega$, which means that due to the back-short and the substrate, absorption of 100 % with a much lower sheet resistance can be achieved compared to a simple sheet resistance in free space. We will see later, that this also narrows the frequency bandwidth down, in which photons are absorbed reducing the optical efficiency.

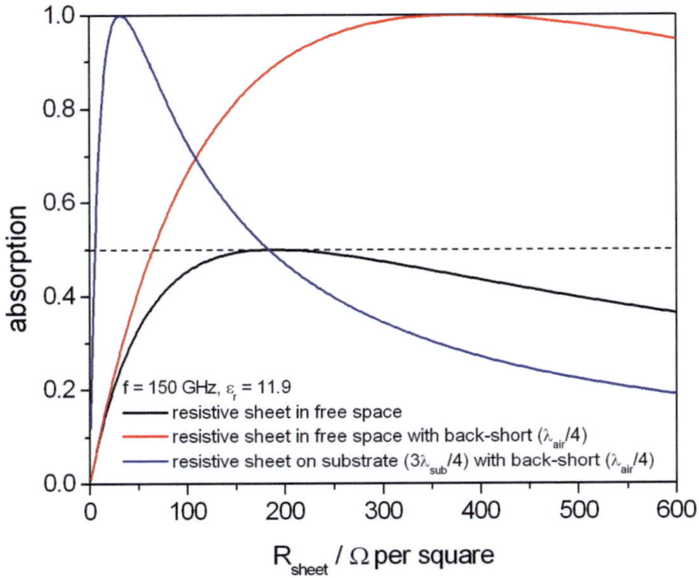

Fig. 7.5.: Comparison of the sheet resistance dependent absorption for three different circuit configurations. black: resistive sheet in free space, red: resistive sheet in free space with back-short; blue: resistive sheet on silicon substrate with back-short.

In Fig. 7.6 the principle of propagation of a mm-wave signal at 150 GHz in a configuration as shown in Fig. 7.4 is demonstrated. The wavelength of the incoming signal is 2 mm (150 GHz). The dielectric constant of the substrate shrinks the wavelength before the signal passes the absorbing resistive sheet. Depending on the absorption, the amplitude of the wave that passes the absorber is reduced. As demonstrated before, the maximum absorption can be achieved for a substrate thickness of $(1+2n)\lambda_{sub}/4$ and a back-short situated at a distance of $(1+2n)\lambda_0/4$.

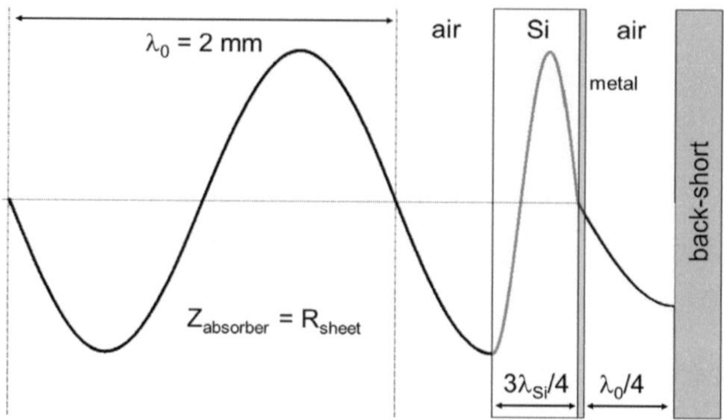

Fig. 7.6.: Propagation of a electromagnetic mm-wave ($\phi = -90\ deg$) at 150 GHz in a one port configuration with substrate, absorbing sheet resistance and back-short cavity.

7.1.4. Optical coupling of the partially filled LEKID absorbing area

In the LEKID case we do not have a solid resistive sheet but some kind of inductive strip grating for a incoming electromagnetic wave. The LEKID meander can be considered as a distributed absorber with a reduced filling factor, where the impedance of the absorbing area depends on the sheet resistance $R_\square$ of the material and the filling factor [61]. If the distance between the parallel meander lines is much smaller than the wavelength, the LEKID meander can be considered as infinite long

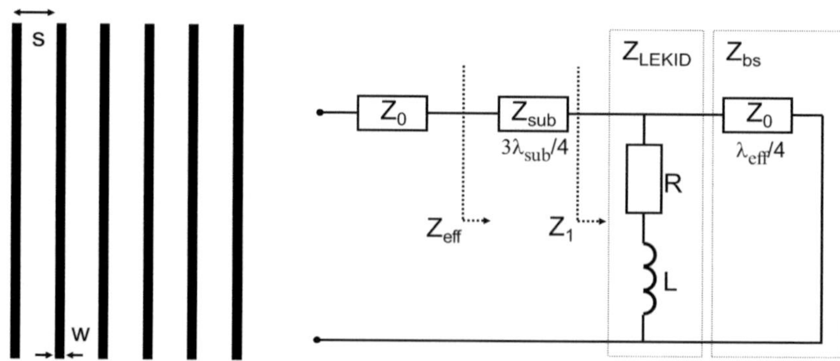

Fig. 7.7.: Left hand side: strip grating of infinite extended parallel lines defined by the filling factor w/s. right hand side: equivalent transmission line model of LEKIDs on substrate with back-short cavity.

parallel strips, where the filling factor is defined by the width w and the spacing s between the lines. This is shown in Fig. 7.7. The impedance of such a strip grating structure is given by [61, 62] and can be calculated with

$$Z_{LEKID} = R + j\omega L = \frac{R_{\square}}{\frac{w}{s}} + jZ_0 \frac{s}{\lambda} ln\ csc\left(\frac{\pi w}{2s}\right) \tag{7.13}$$

where w is the width of the strip lines and s the spacing between them. This equation is only valid for gratings with w $<< s$ and $s < \lambda$ and for a plane wave with a electrical field orientation parallel to the strip lines. For mm-waves with an electrical field perpendicular to the meander lines, the structure is transparent and there is almost no absorption. For the following calculations we will therefore always consider a plane wave with E-field vector parallel to the strip grating.
The impedance of a LEKID is not purely ohmic but has also an inductive reactance ωL due to its partially space filling geometry. The inductance L in 7.13 is a pure geometric inductance. Since the frequency of 150 GHz is higher than the gap energy of the superconductor, we can assume the aluminum as normal conductive [16]. This changes the equivalent circuit discussed before. We have to replace the purely ohmic resistance $R_{\square}$ by the complex impedance Z_{LEKID} (see Fig. 7.7). Z_{LEKID} depends on the sheet resistance and on the filling factor defined as w/s and also on the frequency. The reactance of Z_{LEKID} is independent of the dielectric boundaries because it is a purely inductive circuit [66] considering the meander structure as strip grating. The resistance is already lowered by the $\lambda/4$-transformation, but is still high compared to the sheet resistance of a 40nm aluminum film at low temperatures, which is around 0.5 $\Omega/\square$. The dependency of R on the filling factor is therefore a big advantage and the reason for using this meander geometry because it allows to tune the resistive part to much higher values (see (7.13)). For a LEKID geometry with $s = 4\ \mu m$ and w$= 275\ \mu m$, which corresponds to $w/s = (69)^{-1}$, for example, a resistance $R = 34\ \Omega$ can be achieved.
Due to the complex Z_{LEKID}, the impedance matching as it was shown before has to be modified for the new circuit. The strip grating adds an inductance to the equivalent circuit and effectively shortens the electrical length of the back-short. A shorter back-short ($l_{bs} < \lambda_0/4$) leads to an impedance mismatch and therefore to a lower absorption. For that reason, the back-short distance l_{bs} has to be adapted to this ad-

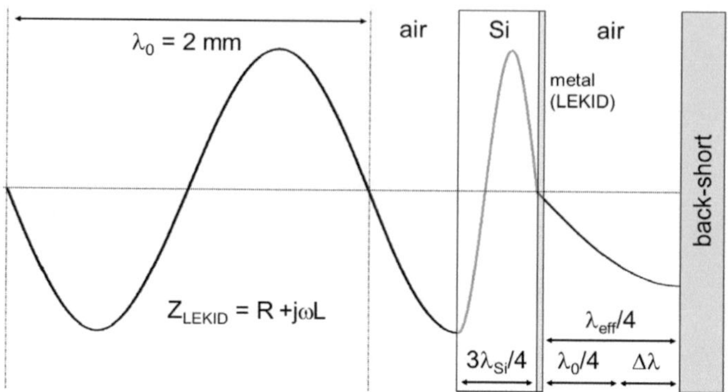

Fig. 7.8.: Propagation of a electromagnetic mm-wave ($\phi = -90\ deg$) at 150 GHz in a one port configuration with substrate, absorbing strip grating and back-short cavity. The back-short has to be adapted by the value $\Delta\lambda$ to compensate the inductance L of Z_{LEKID}.

ditional inductive reactance by increasing the distance to $l_{bs} = \lambda_{eff}/4$. The absorption for this circuit can be calculated by using the same equations as for a resistive sheet on a substrate with back-short (see section 7.1.3), but $R_{\square}$ has to be replaced by Z_{LEKID}. This is demonstrated in the schematic of Fig. 7.8. The incoming electromagnetic wave with $\lambda_0 = 2$ mm passes the substrate, where the wavelength is shrunk to λ_{sub}. Due to the inductance in Z_{LEKID}, the back-short has to be extended by the equivalent electrical length $\Delta\lambda$ of the inductance L to achieve maximum absorption.

For the case of a resistive sheet, $R_{\square}$ has to be matched to the environmental network but is frequency independent. The frequency dependency of Z_{LEKID} makes the impedance matching more complicated, because it is not the same for all frequencies over a relatively large bandwidth. For the NIKA 2.05 mm band (125-175 GHz), the absorption has to be sufficient high over the whole band. Therefore we will now consider the frequency dependent absorption of LEKIDs for the mentioned frequency band. The one port circuit in Fig. 7.7 can be considered as an absorbing transmission line resonator, where the resonance frequency and the amplitude of the absorption depend on the length of the back-short cavity, the substrate thickness and the impedance of the LEKID geometry Z_{LEKID}. Due to the frequency dependence of the absorption, we will optimize the optical coupling for a maximum absorption at the center frequency (150 GHz) with a bandwidth that is as large as possible.

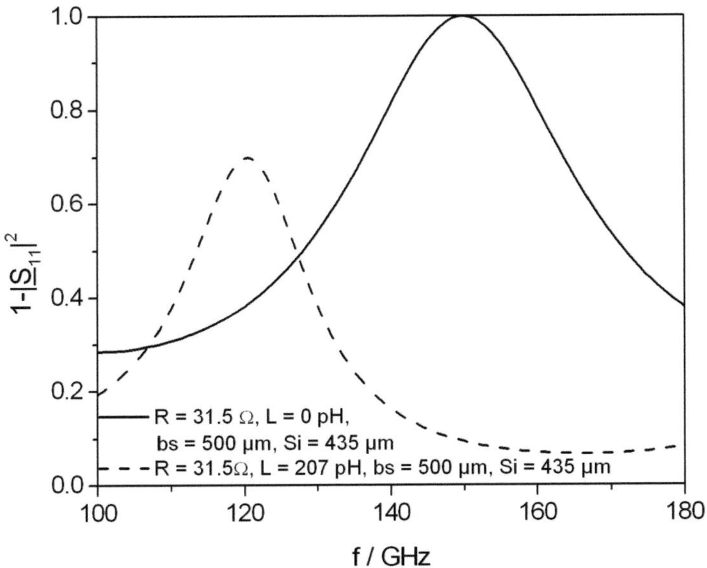

Fig. 7.9.: Influence of the complex impedance on the frequency dependent absorption of LEKIDs. Dashed line: pure ohmic impedance Z_{LEKID}, solid line: complex impedance with inductive reactance. Back-short and substrate thickness have been kept unchanged.

Several calculations have been done to show the influence of the inductance L, the substrate thickness, the back-short and the resistive part R to the frequency dependent absorption. Fig. 7.9 shows the influence of the inductance L on the absorbed frequency band. If $L = 0$ (solid line), the impedance is purely ohmic and the band of absorption is centered at 150 GHz for $l_{sub} = 3\lambda_{sub}/4$ ($435\ \mu m\ for\ silicon$) and $l_{bs} = \lambda_0/4$ ($500\ \mu m$) with $R = 31.5\ \Omega$ as calculated before. Adding now an inductance $L = 207\ pH$, which corresponds to $w/s = (69)^{-1}$, the frequency band, where power is absorbed in the structure, shifts to lower frequencies (center at 120 GHz). Also the band is narrowed down and the maximum absorption decreases. This frequency shift can now be compensated by adapting the substrate thickness and the back-short distance.

The NIKA LEKIDs are fabricated on 300 μm thick silicon substrates for simple practical fabrication reasons and will be kept unchanged from now on. The influ-

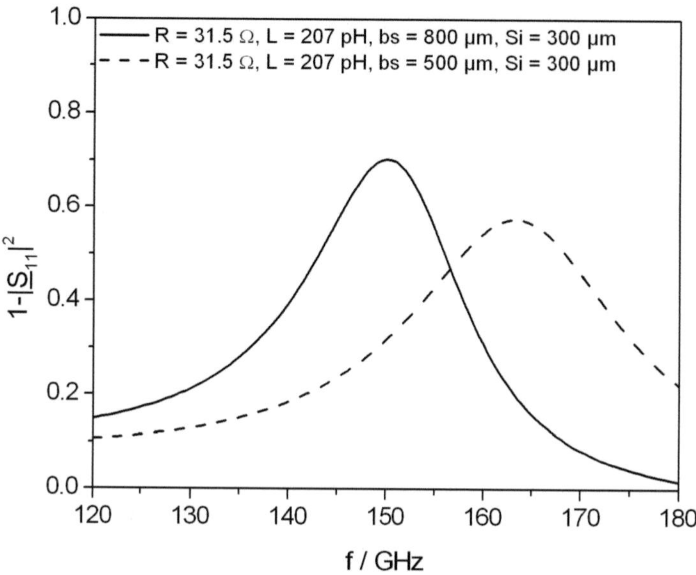

Fig. 7.10.: Influence of the substrate thickness and the back-short to the frequency dependent absorption of LEKIDs. Dashed line: $l_{bs} = 500\ \mu m$, solid line: $l_{bs} = 800\ \mu m$. Z_{LEKID} and the substrate thickness (300 μm) have been kept unchanged.

ence of the changed substrate thickness is demonstrated in Fig. 7.10. The dashed line shows the absorption for a substrate thickness of $l_{sub} = 300\ \mu m$. Compared to the dashed line in Fig. 7.9, the thinner substrate shifts the frequency band to higher frequencies (center at 163 GHz). To re-center the band at 150 GHz for $l_{sub} = 300\ \mu m$, the back-short has to be extended from $l_{bs} = 500\ \mu m$ to $l_{bs} = 800\ \mu m$ (solid line). This means, that even for a substrate thickness $l_{sub} \neq (1+2n)\lambda_{sub}/4$, the mismatch can be compensated in parts by adapting the back-short length, which is of course is a big advantage for determining and optimizing the optical coupling of LEKIDs. The fourth parameter that has big influence on the absorption is of course the resistance R, where the power is actually absorbed. Fig. 7.11 shows the absorption for three different values for R. R is frequency independent and has almost no influence on the position of the absorbed band in the spectrum, but acts on the bandwidth and on the absolute absorption. As already seen, the optimal value for R is not anymore 31.5 Ω. Due to the complex Z_{LEKID} and the thinner substrate, it is more complica-

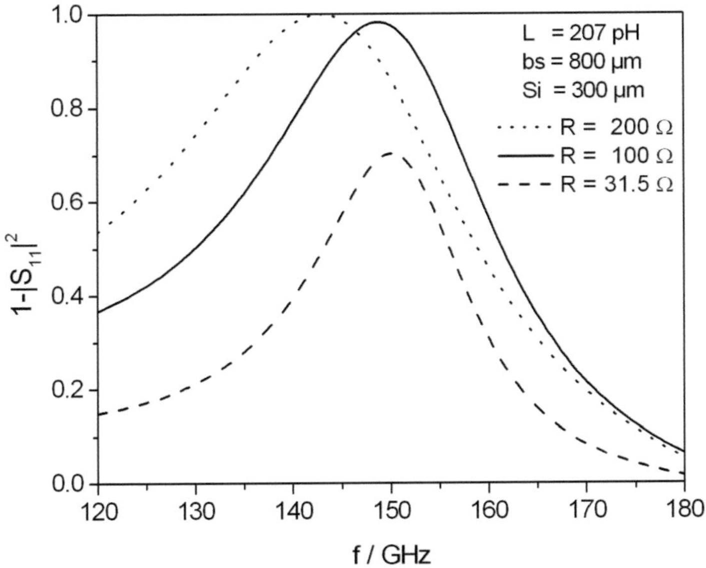

Fig. 7.11.: Influence of R on the frequency dependent absorption of LEKIDs. Dashed line: $R = 31.5\ \Omega$, solid line: $R = 100\ \Omega$, dotted line: $R = 200\ \Omega$. The inductance L, the substrate thickness and the back-short have been kept unchanged.

ted to determine a value R for a impedance matching over a wide frequency band. For each change in impedance, the back-short has to be adapted to achieve not only a center frequency at 150 GHz, but also a reasonable absorption bandwidth. We can obtain the highest absorption for a resistance R, where Z_{eff} is near Z_0 (see equation 7.12). In that case, when $S_{11} = 0$, all power is absorbed, but only in a relatively small bandwidth. Therefore, a good compromise between absolute absorption and sufficient large bandwidth has to be found. The example in Fig. 7.11 (solid line) with $R = 100\ \Omega$ and $w/s = (69)^{-1}$ gives already a reasonable absorption and is a solution that is realizable. A corresponding LEKID geometry would be w$= 4\ \mu m$, $s = 276\ \mu m$ and $R_{\Box} = 1.45\ \Omega/square$. Sheet resistances in that range are possible with an aluminum film thickness of around 20 nm at low temperatures.

We have shown here the possibility of modeling and optimizing the optical coupling of LEKIDs with a relatively simple transmission line model. As we will see later, the model agrees very well with measurements of such structures and can therefore be used as a first optimization tool for the optical coupling of these meander line LEKIDs.

7.2. Reflection measurements at room temperature

The calculation with the transmission line model shown above allows to model the optical coupling by considering the LEKID as inductive parallel strips. This does not take into account the structure of the interdigital capacitor or the ground plane around the pixels, which have certainly less influence on the total absorption but cannot be completely neglected. To verify the results from the calculations and to check the influence of these parameters, the optical absorption of mm-waves in LEKIDs has been measured. Cryogenic measurements are very time consuming especially if a large number of series of samples has to be measured. For each sample it would take a lot of time for the cool down and the heat up process. For this reason we use a setup that allows measurements at room temperature, whose results are comparable to the cryogenic case. A measurement setup of this type and the results will be demonstrated and discussed in this section. The results will be compared to calculations done with the transmission line model and full wave simulations performed with CST Microwave Studio [41].

7.2.1. Measurement setup

Rf-setup One possibility to measure the optical coupling of LEKID structures is with a reflection measurement setup that is comparable to the real detection principle and that corresponds to the detection principle shown in Fig. 7.12. To perform such measurements, a relatively complex rf-setup is required to measure the reflection parameter S_{11} over a large bandwidth and at such high frequencies (120-180 GHz) with a sufficient high dynamic range. A schematic of such a setup is shown in Fig. 7.12 including the quasi-optical part and the sample. The source output of the network analyzer generates a frequency band from 20 to 30 GHz. This band is mixed up to 120 to 180 GHz using a harmonic mixer, which provides the frequencies of the 2mm band. One part of the signal is mixed down with a local oscillator signal to IF_{ref} and serves as a reference signal. The second part is guided to the feed horn. The signal that is reflected at the back-short and not absorbed in the sample is collected by the same feed horn. It is mixed down with the same local oscillator

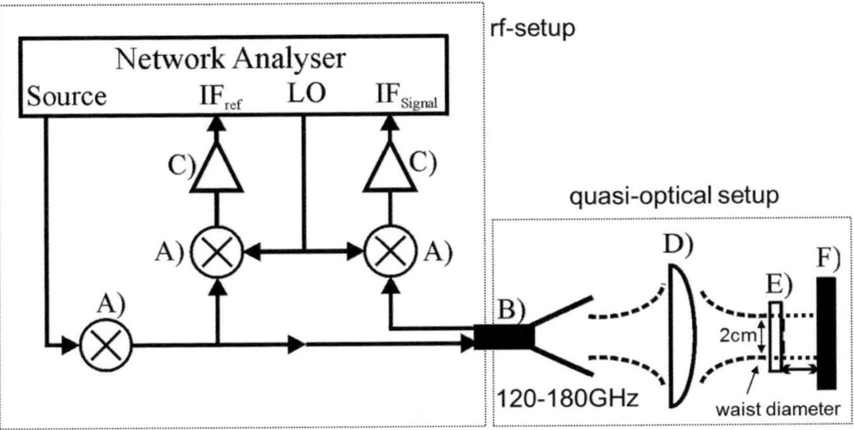

Fig. 7.12.: Reflection measurement setup to measure the absorbed power in LEKID structures at room temperature. A) Harmonic mixer, B) Corrugated feed horn, C) Amplifier, D) Corrugated quasi-optical lens, E) Sample, F) Back-short with variable distance to sample.

signal to IF_{Signal}. In the network analyzer the reflected signal IF_{Signal} is compared to IF_{ref} and S_{11} is determined. More details about this rf-setup can be found in [36] allowing the calculation of the absorption using (7.8).

This measurement principle assumes that all absorbed power in the LEKID is considered as detected power. To be comparable to the low temperature measurements, the sheet resistance $R_\square$ of the aluminum film has to be adapted to that at room temperature. As mentioned before, for frequencies higher than the gap frequency of aluminum, it can be considered as normal conductive for an incoming electromagnetic wave. The normal resistivity at low temperatures is lower than that at room temperature. The factor in resistivity is usually described by the residual resistance ratio (RRR) (see Chapter 5). The resistivity or the sheet resistance can therefore be adapted by the film thickness. An aluminum film with $R_\square$=0.5 $\Omega/square$ at low temperature and $R_\square$=1.5 $\Omega/square$ at room temperature has an $RRR = 3$. To achieve a room temperature sheet resistance of 0.5 $\Omega/square$, the film has to be made three times thicker. Some $R_\square$-values for different film thicknesses and temperatures, that have been measured, are shown in table 5.1.

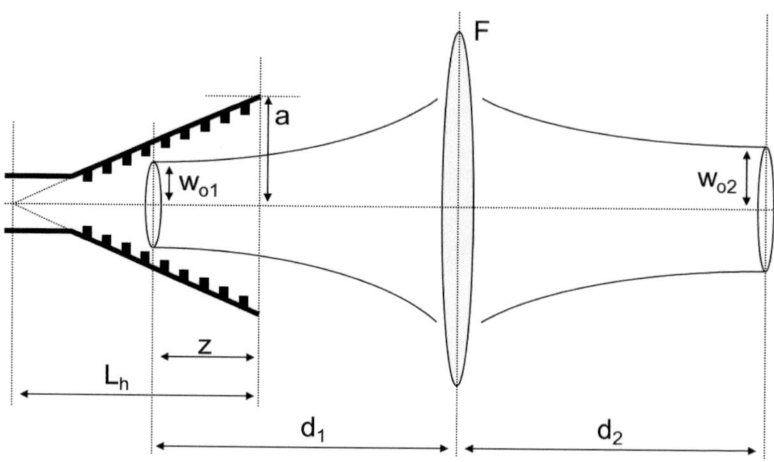

Fig. 7.13.: Quasi-optical part of the reflection measurement setup.

Quasi-optical setup In the Fig. 7.12 a corrugated feed horn is used as *waveguide-to-Gaussian beam* converter, or in other words, as Gaussian beam launcher. Corrugated feed horns possess launching efficiencies of around 98 % with a wide bandwidth and excellent cross-polarization. The supressed cross-polarization allows an almost perfect plane wave with only one polarization. But it leads also to the fact that reflected power in different polarizations, is not detected by the feed horn and is considered as absorbed power. In Fig. 7.13 the quasi-optical setup is shown, consisting of the feed horn and a quasi-optical corrugated lens. The lens allows to refocus the Gaussian beam on the sample. The size of the beam, at the position, where the sample is mounted, is given by the so-called minimum waist w_{o2} and depends on the waist of the feed horn w_{o1}, d_1, d_2 and the focal length L_h of the lens. w_{02} should not exceed the size of the sample in order to avoid reflections on the sample holder around it. We have chosen a waist size of $\mathrm{w}_{o2} = 10$ mm in order to keep the sample size relatively small with maximum side length of 25 mm. To achieve that, d_1, d_2 and the waist of the feed horn w_{o1} have to be calculated. w_{o1} depends on the length L_h of the feed horn and the frequency. From [19] it can be calculated with

$$\mathrm{w}_{o1} = \frac{\tau a}{\sqrt{1+(M_h\gamma^2)^2}} \tag{7.14}$$

where

$$M_h = \frac{2\pi a}{\lambda} \frac{a}{L_h + \sqrt{L_h^2 + a^2}} \tag{7.15}$$

with a the radius at the mouth of the feed horn and $\gamma \approx 0.6435$. γ is defined as the ratio of w_{o1}/a at which the normalized coupling coefficient is 0.99 and 98 % of the power is contained in the fundamental Gaussian beam mode [26]. To determine d_1, the waist position in the feed horn has to be known and is given by

$$z = L_h \left(1 - \frac{1}{1 + (M_h \gamma^2)^2}\right) \tag{7.16}$$

The used corrugated lens is optimized for 150 GHz with a focal length of $F = 55$ mm. The most homogeneous and focused plane wave in the setup can be found at the distance d_2, where the waist is at its minimum (w_{o2}). For the used feed horn with $L_h = 55$ mm and $a = 5.5$ mm, we get $w_{o1} = 3.3$ mm and $z = 6.2$ mm. w_{o2} is then given by [19]

$$\mathrm{w}_{o2} = \frac{\mathrm{w}_{o1}}{\sqrt{(\frac{d_1}{F} - 1)^2 + (\frac{z_{c1}}{F})^2}} \tag{7.17}$$

where z_{c1} is the confocal distance, which is defined as

$$z_{c1} = \frac{\pi \mathrm{w}_{o1}^2}{\lambda} \tag{7.18}$$

With $F = 55$ mm it is now possible to calculate d_1 for which $\mathrm{w}_{02} = 10$ mm. By solving (7.17) we calculate $d_1 = 55$ mm. Now we can determine the distance d_2, where the sample has to be mounted, which is given by

$$d_2 = F \left(1 + \frac{\frac{d_1}{F} - 1}{(\frac{d_1}{F} - 1)^2 + (\frac{z_{c1}}{F})^2}\right) \tag{7.19}$$

For our setup, we calculate $d_2 = 55$ mm for a minimum waist of $\mathrm{w}_{o2} \approx 10$ mm.

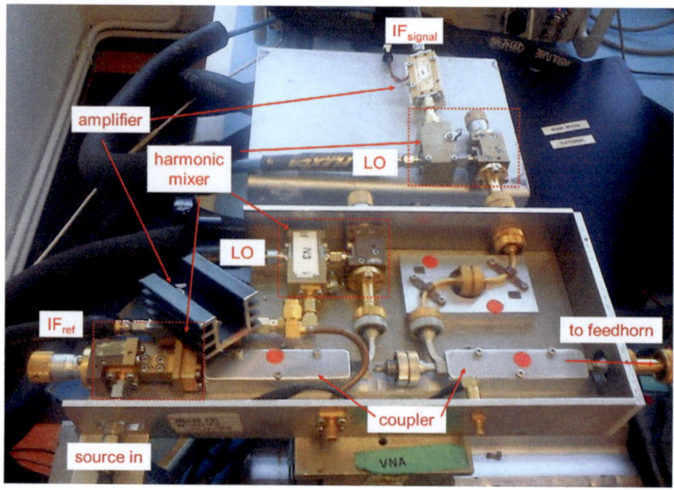

Fig. 7.14.: Picture of the rf-part of the reflection measurement setup.

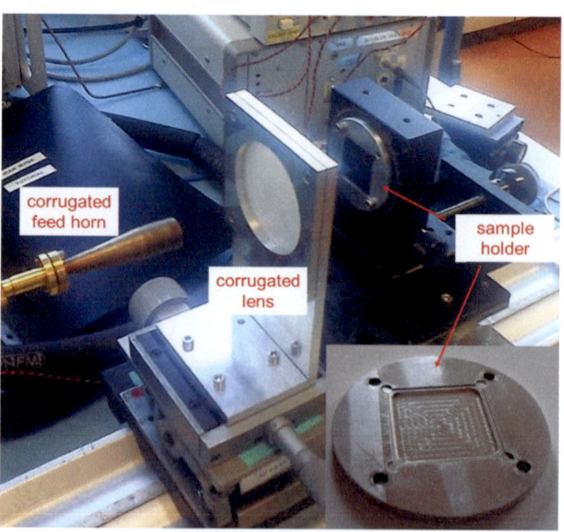

Fig. 7.15.: Picture of the quasi-optical part of the reflection measurement setup including the feed horn, the lens and the mounted sample. The aluminum sample holder with integrated back-short is enlarged on the right bottom side.

In Fig 7.14 and 7.15 two pictures of the measurement setup are shown. One shows the rf-setup with all its components and the second one the quasi-optical setup with the mounted sample. The sample holder is enlarged on the right bottom side and consists of a single aluminum piece combining the holder and the back-short. Different sample holders for different back-shorts have hereby been fabricated. The difficulty with this quasi-optical setup is the precision that is needed by mounting the setup. If the different parts are not perfectly aligned to each other, this will lead to errors in the measurement results. If there is power reflected to the side of the lens due to a bad angle, it is not detected by the feed horn and is considered as absorbed power. High precision aligning elements have therefore been used.

Sample The sample itself consists of an array of 81 LEKIDs (9x9) designed and fabricated for these measurements. The pixels are fabricated on a high resistivity silicon substrate to reduce dielectric losses, each surrounded by a ground plane frame to take its influence on the total absorption into account. The higher the resistivity, the lower the losses and the more transparent the substrate for electromagnetic waves at around 150 GHz. A schematic of such an array, but with only 16 LEKIDs, is shown in Fig. 7.16. The sample has a size of 25x25 mm^2 and is slightly bigger than $2 \cdot w_{o2}$, to avoid reflections on the sample holder and to allow a proper calibration

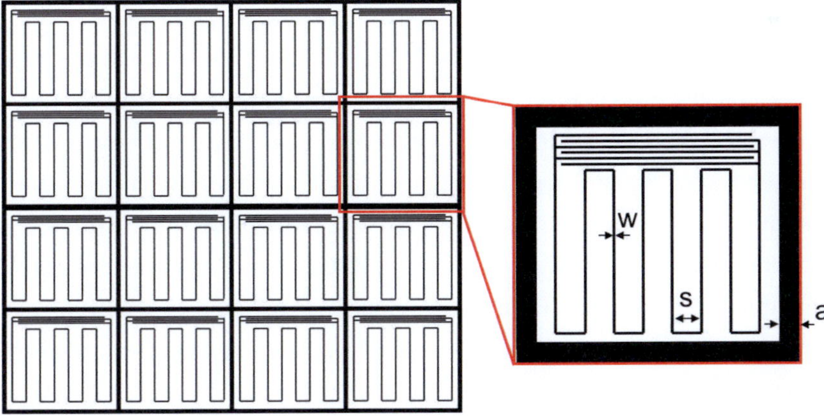

Fig. 7.16.: Schematic of the 9x9 mm sample to measure the optical coupling of LEKIDs. For better visibility, only 16 pixels of the array are shown and the dimensions are exaggerated. Enlarged: single pixel geometry.

of the setup. The measurement of a single LEKID with a size of around 2x2 mm^2 would make the quasi-optical setup much more complicated due to a waist size of half the LEKID size. The array is therefore designed as a periodic structure of many LEKIDs, where the absorption of the whole array can directly be compared to a single LEKID. The LEKIDs in these test samples have thus all exactly the same geometry in order to guarantee the periodicity of the structures.

Possible measurement errors There are several effects that can lead to errors by calculating the absorption from the measured S_{11} with this setup. As mentioned before, the feed horn can only detect signals in one polarization and reflected power in different polarizations due to the pixel geometry will thus not be taken into account and is considered as absorbed. The same problem occurs in the case of side lobe effects or if higher modes than the fundamental mode are excited in the structure. Both are considered as absorbed and lead to higher estimations of the absolute absorption values. These effects are assumed to be small due to the polarized geometry of the LEKID meander, which dominates the absorption. The lines of the interdigital capacitor are perpendicular to the electrical field of the incoming plane wave and the absorption in this part is therefore small, which also reduces the amount of power reflected in other polarizations.

Calibration The system has been calibrated in the back-short plane using a 3-point calibration (short, offset-short, load). For the short and offset-short we used the sample holder shown in Fig. 7.15 without the test sample to calibrate in the back-short plane. For the load we used an absorber material holding it in the Gaussian beam path between the lens and the sample holder. The calibration has been done over the frequency band of 120-180 GHz. Due to instabilities in dynamic range of the harmonic mixers, the calibration at 170-180 GHz could not be done correctly. This might lead to parasitic effects and a comparison to simulation and calculation might be more difficult in this frequency range. For each new back-short, the setup has been re-calibrated. The resulting dynamic range of the system after calibration is around -50 dB over almost the whole frequency bandwidth, which is sufficient for this type of measurement since $S_{11} = -20$ dB already leads to an absorption of 99 % (see equation (7.8)).

7.2.2. Reflection measurement results for different LEKID geometries

The measurements that are presented in this section have been done with 3 different LEKID geometries ($LEKID1,2,3$). The different parameters of these three samples are the square resistance $R_\square$, the meander filling factor ff and the width of the ground plane a, which are listed in Table 7.1. The table also shows the resulting values of R and L forming the mentioned grid impedance Z_{LEKID}. $LEKID1$ has a relatively low $R_\square$ and filling factor (ff), but the surrounding ground plane is much wider than for the other samples. $LEKID2$ has the same filling factor, but higher $R_\square$ and a narrower ground plane frame. $LEKID3$ has a higher filling factor, leading to a higher R and L. The ground plane is of the same dimensions than for $LEKID2$. In the following, the back-short dependent absorption measurement results for these three samples are shown and discussed. The sample geometries have been chosen to investigate the influence of ground plane, filling factor and material resistivity on the LEKID absorption propertiesand to compare it to the calculations shown above. The impact of the resistivity and the filling factor on the absorption has already been demonstrated by calculations but the ground plane influence has not been determined so far.

	$R_\square/\Omega/\square$	$w/\mu m$	$s/\mu m$	$ff^{-1} = s/w$	R/Ω	L/pH	$a/\mu m$
LEKID1	0.65	4	276	$(69)^{-1}$	45	207	400
LEKID2	1.3	4	276	$(69)^{-1}$	90	207	160
LEKID3	1.3	3	276	$(93)^{-1}$	120	224	160

Table 7.1.: Optical coupling parameters of the different LEKID geometries measured with the reflection setup.

LEKID1 Fig. 7.17 shows the back-short dependent absorption calculated from the measured S_{11} of *LEKID*1. As expected, the frequency band, where power is absorbed in the LEKID, shifts to higher frequencies for a decreasing back-short length. This agrees with what we calculated with the transmission line model (comparing Fig. 7.10). The shape of the curve is, on the other hand, different from the calculations. There are two resonance peaks at 125 and 170 GHz and some kind of plateau around 145 GHz. The height of the plateau is decreasing with decreasing back-short. These effects do not appear with the transmission line model and have therefore to be caused by something that is not considered in the calculations such as the ground plane frame around the pixels or the interdigital capacitor.

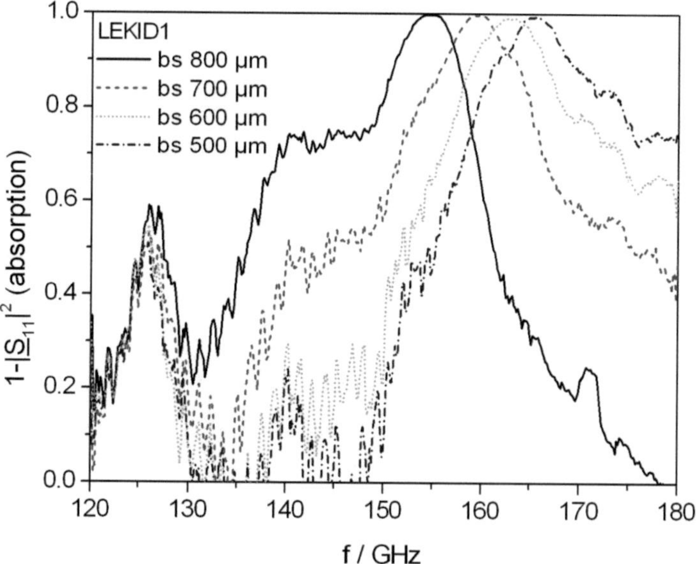

Fig. 7.17.: Back-short dependent mm-wave absorption of *LEKID*1 calculated from the measured S_{11}.

LEKID2 The measurement of the second sample ($LEKID2$) is shown in Fig. 7.18. For a back-short distance of $800\mu m$, the band of absorption is well centered at 150 GHz and the absorption is higher than 50 % over almost 30 GHz of the total bandwidth. This is already an excellent result for a direct detector. The parasitic effects seen from the previous sample are much smaller or not existing at all. This measurement of $LEKID2$ agrees very well with the calculations of the transmission line model in Fig. 7.11 (solid line).

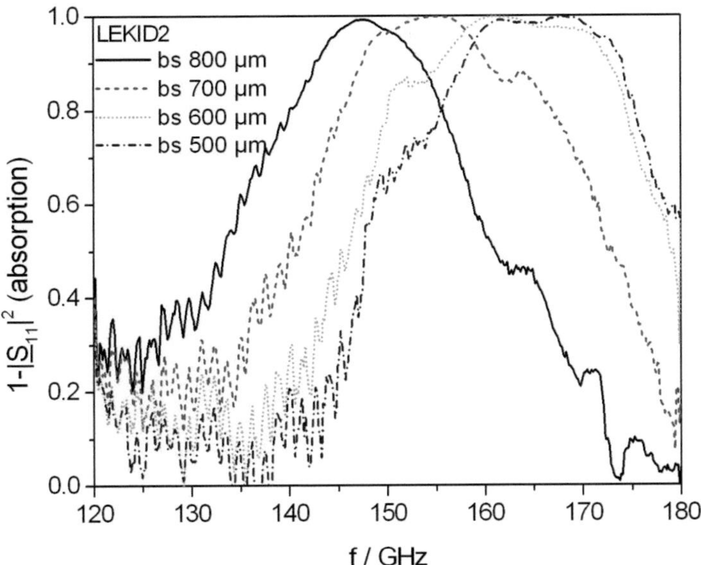

Fig. 7.18.: Back-short dependent mm-wave absorption of $LEKID2$ calculated from the measured S_{11}.

LEKID3 To increase the absorption even more, the filling factor for *LEKID*3 has been decreased to achieve a higher resistance *R*. The result of the reflection measurements for different back-short lengths is shown in Fig. 7.19. Because of the higher inductance *L*, due to the narrower meander lines, the optimal back-short length is at $700\mu m$, where the absorbed band is centered at around 150 GHz. Compared to *LEKID*2, the absorption is slightly higher over a larger frequency band. It is higher than 50 % over more than 30 GHz and confirms the calculations done with the transmission line model.

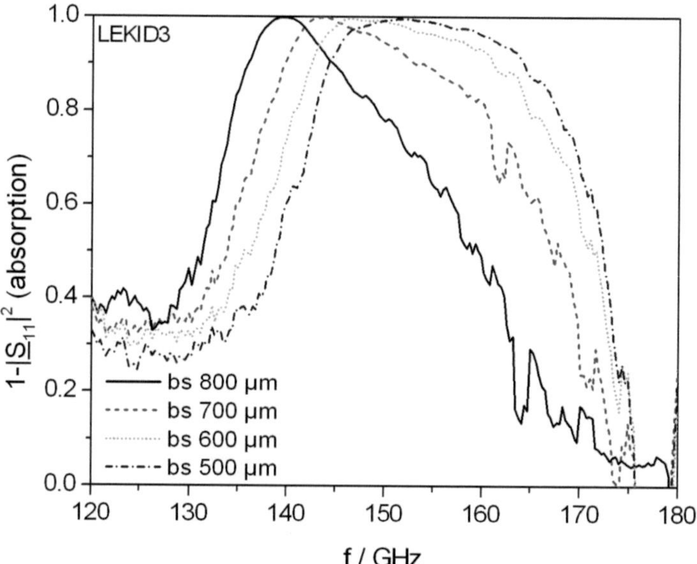

Fig. 7.19.: Back-short dependent mm-wave absorption of *LEKID*3 calculated from the measured S_{11}.

Comparison In Fig. 7.20 the absorption of the three measured samples for the optimal back-short length (center frequency at around 150 GHz) are compared. For $LEKID1$ and 2 it was found to be at $l_{bs} = 800\ \mu m$ and for $LEKID3$ at $l_{bs} = 700\ \mu m$. The higher inductance L of $LEKID3$ leading to a higher impedance Z_{LEKID}, shortens the optimal back-short length to 700 μm. We see, that we achieve already high absorption for all three geometries. The optimization of the optical coupling by lowering the filling factor and increasing the resistivity of the aluminum by using a thinner film, improves the absorption and corresponds to the calculation done with the transmission line model.

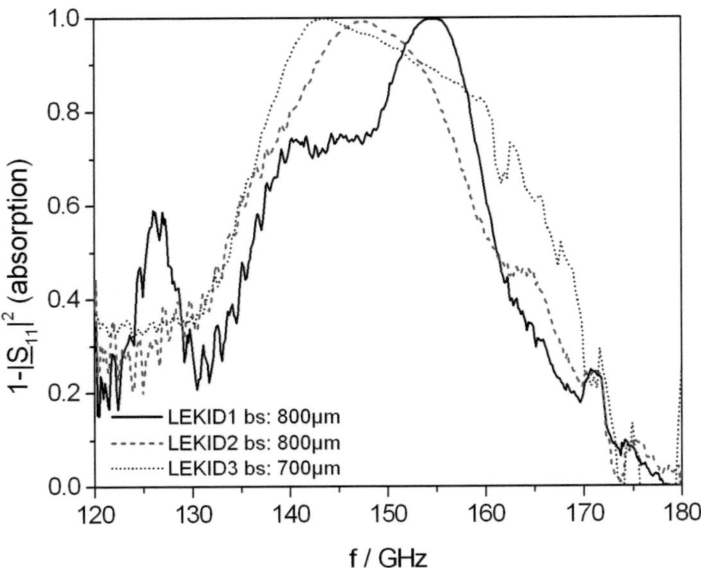

Fig. 7.20.: Comparison of the mm-wave absorption of $LEKID1, 2$ and 3 for optimal back-short length calculated from the measured S_{11}.

7.3. Analysis and interpretation of measurement results

The measurement results seen above can already be well described and explained with the calculations of the transmission line model. But there are still some differences between them, especially for the geometry $LEKID1$. To understand these effects, a way to simulate the entire LEKID structure including the interdigital capacitor and the ground plane frame in order is presented and discussed in this section. The results are compared to the measurements and the transmission line model.

7.3.1. Simulation model

CST Microwave studio offers the possibility to perform full wave simulations of such structures, where we can simulate the absorption of this one port system by modeling it as a rectangular waveguide. A schematic of the CST model is shown in Fig. 7.21 A). At port 1, a wave guide port is placed that generates the incoming mm-wave signal. To achieve a plane wave propagating in this waveguide the boundary conditions have to be defined accordingly, which is demonstrated in Fig. 7.21 B). To generate it with an electrical field parallel to the meander strips of the LEKID, the boundary conditions have to be set as shown. The top and bottom side as electrically equal to zero ($E = 0$) and the back and front side as magnetically equal to zero ($H = 0$). This way, the electrical field is parallel and the magnetic field perpendicular to the long meander lines of the LEKID. The back-short is modeled as an electrically short ended wall with $E = 0$. The aluminum material properties are

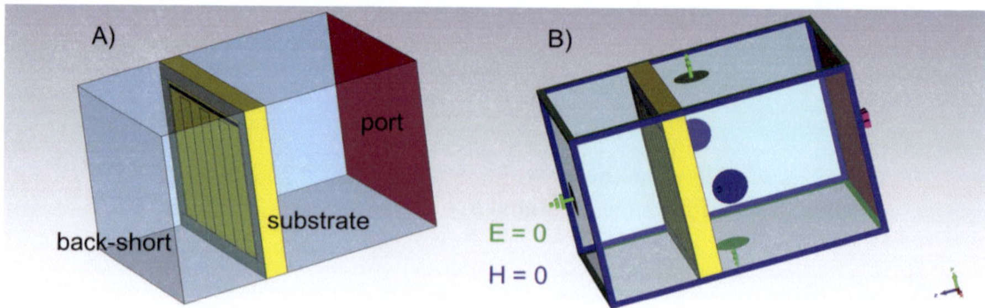

Fig. 7.21.: A) Schematic of the CST model to simulate the optical coupling of LEKIDs. B) The model is considered as a waveguide defined by the boundary conditions $H = 0$ (blue) and $E = 0$ (green). The back-short is modeled as an electrically shorted wall with $E = 0$.

defined by the electrical conductivity and the thickness of the film. The silicon substrate parameters are given by the permittivity ($\varepsilon_r = 11.9$) and the loss tangent. The loss tangent is small due to the high resistivity of the substrate even at room temperature and was assumed to be $tan\ \delta \approx 10^{-4}$. With these parameters, the model is sufficiently defined for simulations that are comparable to the room temperature reflection measurement results.

7.3.2. Comparison of measurement, numerical simulation and transmission line model

LEKID1 The numerical simulations allow to model the exact geometry of the LEKIDs including the interdigital capacitor and the ground plane to explain and to understand the measurement results better. Fig. 7.22 shows the comparison of the measurement of $LEKID1$ compared to the calculation done with the transmission line model (no ground plane, no capacitor) and the full wave simulation (with ca-

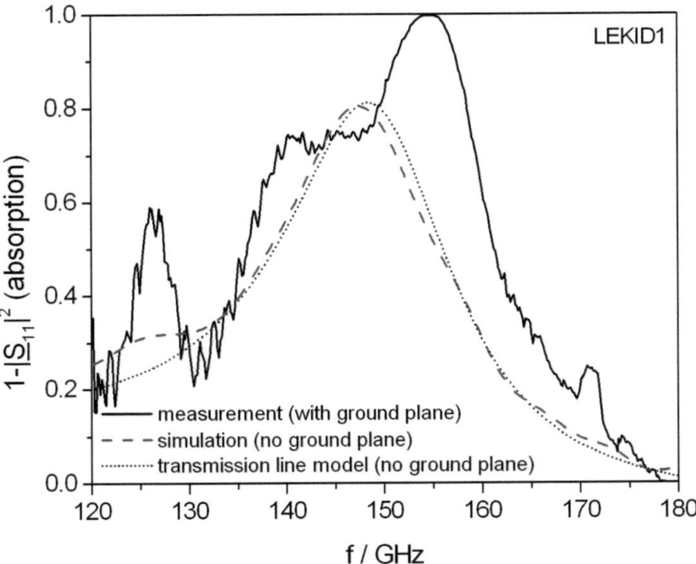

Fig. 7.22.: Numerical simulations (dashed line) and calculation using the transmission line model (dotted line) without ground plane frame compared to the measurement of $LEKID1$ (solid line). The back-short distance is $l_{bs} = 800\ \mu m$.

pacitor, no ground plane). We see that calculation and simulation match almost perfectly, but both are different from the measurement. The good agreement between both shows that the difference to the measurement is not caused by the interdigital capacitor because it is included in the CST-model. Therefore, the assumption of less absorption in the capacitor strips, perpendicular to the electrical field lines, is confirmed. Thus, the reason for the unexpected measured absorption must be the surrounding ground plane, which is neither considered in the calculation nor in the simulation so far. To verify its influence on the total absorption, a second simulation has been done, this time including the entire geometry (meander line, capacitor and ground plane). The result of this simulation is shown and compared to the measurement in Fig. 7.23. The ground plane width of the sample is $a = 400\ \mu m$. For the simulation of only one LEKID structure, with respect to the periodicity of the sample, the ground plane has a width of $a/2 = 200\ \mu m$. Adding the ground plane leads to a good agreement of the measured and simulated $LEKID1$. It is therefore

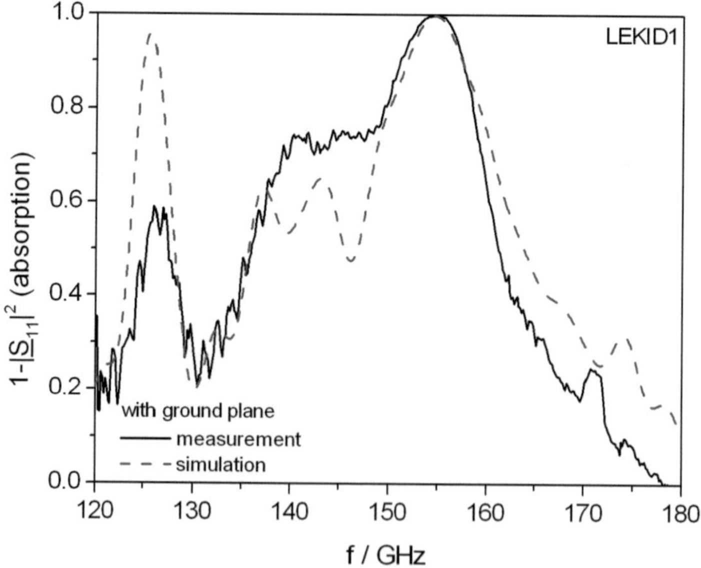

Fig. 7.23.: Numerical simulations with ground plane frame (dashed line) compared to the measurement of $LEKID1$ (solid line). The back-short distance is $l_{bs} = 800\ \mu m$ and the ground plane width $a = 200\ \mu m$.

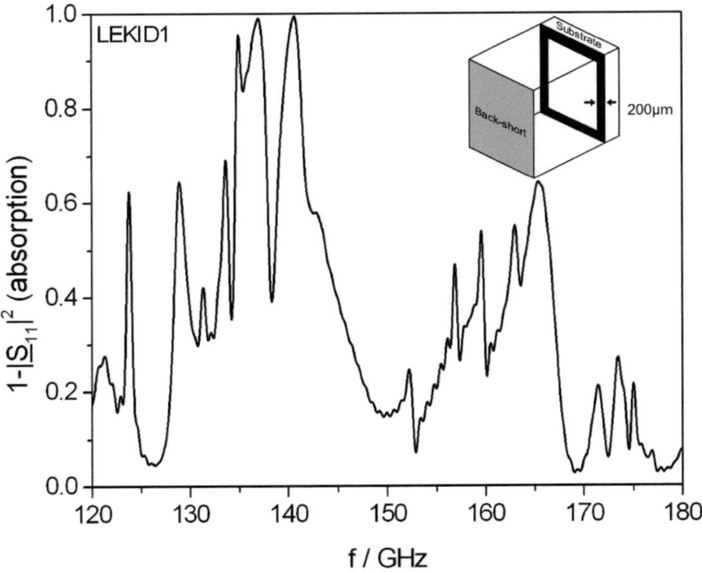

Fig. 7.24.: Numerical simulation of a 200 μm ground plane frame on a silicon substrate without LEKID geometry. The back-short distance is $l_{bs} = 800\ \mu m$.

obvious, that it has a non-negligible influence on the total calculated absorption. Later, when we use the LEKID as detector, only power that is absorbed in the meander can be considered as detected power. This means, that a part of the power might also be absorbed in the ground plane, but cannot be considered as detected. From the measured S_{11} we calculate the total absorption, which includes the power absorbed in the LEKID and in the ground plane. If the influence of ground plane is too high, using the reflection measurement setup, it becomes difficult to determine the real absorption of the meander structure. The ground plane not only changes the pixel geometry but also the size of the pixel. Adding the size of the ground plane to the pixel, changes the total filling factor. The LEKID size is the same in both cases, but the ratio of pixel size and direct absorption area changes. This leads to a change in effective impedance of the geometry and therefore to a different matching between Z_0 and Z_{eff}. The two parts of the pixel, ground plane and LEKID, are not independent from each other, which means that we cannot consider the total impedance as a simple superposition of both individual impedances. The two build a common impedance as a combination of both geometries. We can verify this by simulating the ground plane frame without the LEKID geometry. The result of this is shown in Fig. 7.24. If both structures were independent, the superposition of the simulation in Fig. 7.24 and Fig. 7.22 would result in the simulation of the entire geometry (Fig 7.23). It is obvious, that this is not the case. This is a important information, that the ground plane not only absorbs a part of the power, but also changes the pixel impedance. Comparing the measurement and the simulation of the ground plane frame, similarities can be found, that indicate the influence of the frame. The mentioned plateau at 140 GHz corresponds most likely to the increased absorption in Fig. 7.24 between 130 and 140 GHz. The increased absorption at 125 and 170 GHz can also be re-found in Fig. 7.24 at 123 and 173 GHz.

LEKID2 The ground plane frame of *LEKID*3 is narrower (160 μm) and the influence to the total absorption should be smaller. Therefore, the same simulations have been done as for *LEKID*2 to verify the influence of the frame width. Fig. 7.25 shows the comparison of the measurement (with ground plane), the simulation and calculation both without ground plane. For the smaller frame, the three curves match much better than for the previous sample and indicate less parasitic absorption or impedance transformation effects. Compared to the simulation including the ground plane in Fig. 7.26, the difference in absorption is much smaller, but still existing. Between 120 and 130 GHz, the absorption decreases with the existing ground plane, which confirms the assumption of a combined effective impedance of LEKID and ground plane. For optimal coupling, the back-short length l_{bs} has to be optimized for the whole structure by simulation and the room temperature reflection measurements. The less influence of the ground plane is confirmed by the full wave simulation of a 80 μm wide ground frame without the LEKID geometry. The result

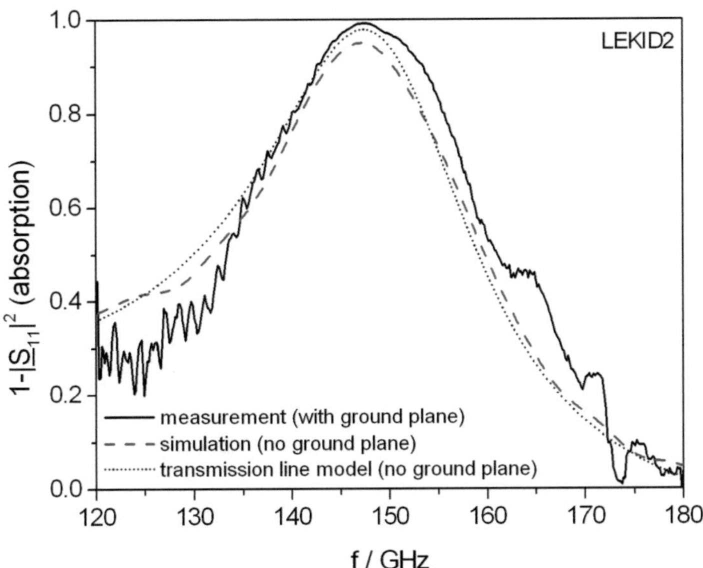

Fig. 7.25.: Numerical simulations (dashed line) and calculation using the transmission line model (dotted line) without ground plane frame compared to the measurement of *LEKID*2 (solid line). The back-short distance is $l_{bs} = 800\ \mu m$.

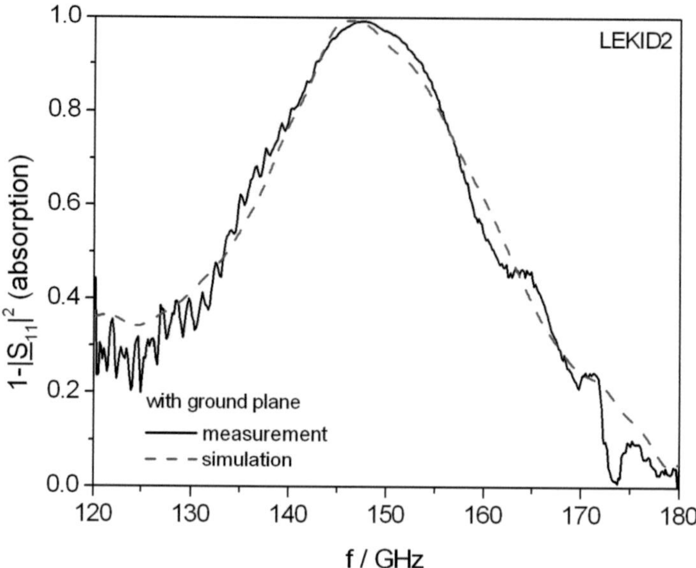

Fig. 7.26.: Numerical simulations with ground plane frame (dashed line) compared to the measurement of *LEKID2* (solid line). The back-short distance is $l_{bs} = 800\ \mu m$ and the ground plane width $a = 80\ \mu m$.

in Fig. 7.27 shows, that the absorption in the narrower frame is much smaller and almost negligible compared to the 200 μm frame of *LEKID2*.

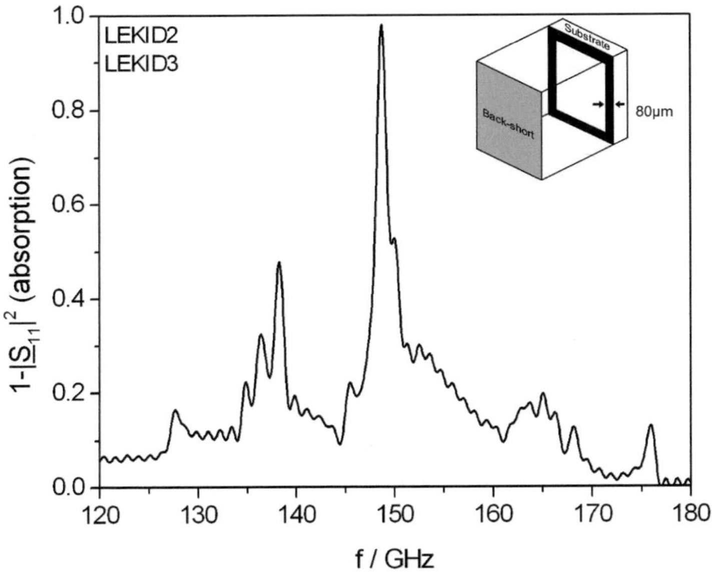

Fig. 7.27.: Numerical simulation of a 80 μm ground plane frame on a silicon substrate without LEKID geometry. The back-short distance is $l_{bs} = 800\ \mu m$.

LEKID3 The highest absorption for *LEKID*3 is achieved with a back-short of $l_{bs} = 700\ \mu m$. This is due to the higher inductance L leading to a shorter back-short for a center frequency at 150 GHz, as mentioned before. As for the two previous samples, the measurement of *LEKID*3 is first compared to the simulation and calculation without ground plane frame (see Fig. 7.28). The back-short is well adapted for a center frequency at 150 GHz in the three cases. The shape of the measured curve agrees less with the simulation and calculation than *LEKID*2. The ground plane width is the same for both geometries, but the back-short length is shorter. The effective impedance of LEKID and ground plane is different, not only because of the higher inductance L and resistance R, but also due to the shorter back-short. This changes the influence of the ground plane for *LEKID*3 compared to *LEKID*2. To verify this effect, *LEKID*3 has also been simulated including the ground plane (see Fig. 7.29). The simulation matches now much better with the measurement, except for frequencies between 170 and 180GHz. This can be explained by the mentioned calibration errors in this frequency range due to the limited dynamic range of the harmonic mixers. This result shows, that the influence of the a identical ground plane frame (as *LEKID2*) depends also on the impedance Z_{LEKID} and the back-short length l_{bs} and not purely on its geometry.

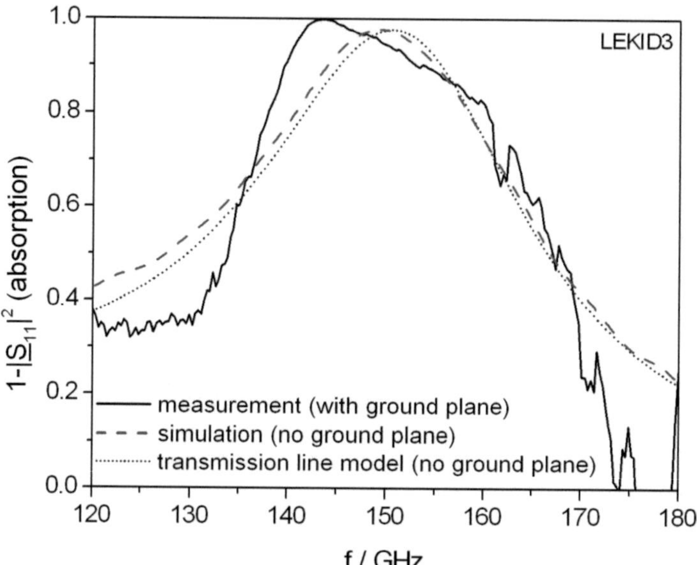

Fig. 7.28.: Numerical simulations (dashed line) and calculation using the transmission line model (dotted line) without ground plane frame compared to the measurement of *LEKID3* (solid line). The back-short distance is $l_{bs} = 700\ \mu m$.

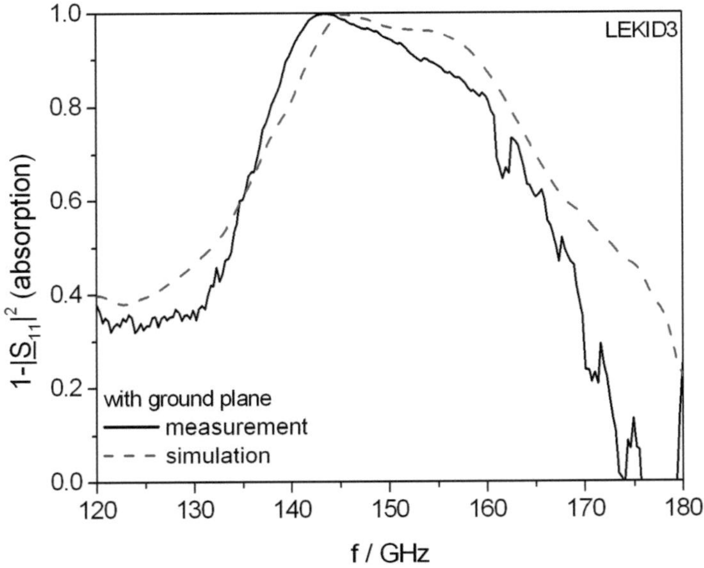

Fig. 7.29.: Numerical simulations with ground plane frame (dashed line) compared to the measurement of *LEKID*3 (solid line). The back-short distance is $l_{bs} = 700\ \mu m$ and the ground plane width $a = 80\ \mu m$.

Comparing the mm-wave coupling of $LEKID1,2,3$ **without ground plane** The three structures showed different dependencies in optical coupling due to different ground planes around the pixels. But also the absolute absorption in the meander structure, not considering the ground plane, varies due to different values of the sheet resistance $R_\square$ already seen above. To demonstrate the improvement in total absorption dependent of $R_\square$, Fig. 7.30 shows the comparison of CST simulations of the three geometries ($LEKID1,2,3$) without the ground plane. Comparing the optical coupling of the meander section independent of the ground plane demonstrates the improvement due to a different filling factor and resistivity even better. The narrower lines of w$=$ 3 μm and the higher square resistance $R_\square = 1.3\ \Omega/\square$ of $LEKID3$ result in a much higher absorption over the bandwidth of 120-180 GHz compared to $LEKID1$ (w$=$ 4 μm and $R_\square = 0.65\ \Omega/\square$).

With this optimized optical coupling, that allows absorption of up to 100 % at the maximum, we now have a detector geometry that is highly sensitive to mm-waves

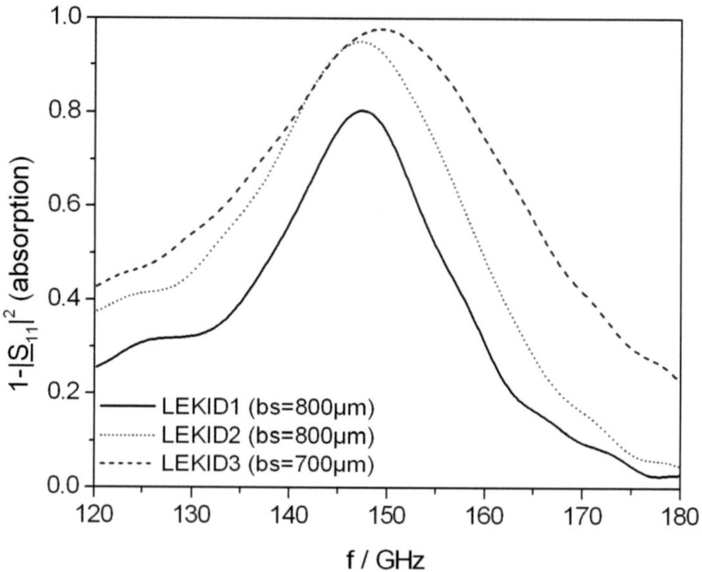

Fig. 7.30.: Comparison of numerical simulations for the three geometries $LEKID1,2,3$ without ground plane frame. The back-short length for $LEKID1$ and 2 is $l_{bs} = 800\ \mu m$ and for $LEKID3$ $l_{bs} = 700\ \mu m$.

in the band of 120-180 GHz. Based on these results, arrays of 132 pixels have been designed and fabricated to determine the electrical and optical properties of large LEKID arrays at low temperatures. This characterization will be discussed in the following chapter.

8. LEKID characterization in the NIKA camera

After the investigation of electrical and mm-wave absorption LEKID parameters for the 2.05 mm band as described in the previous chapters, we designed and fabricated arrays of 132 pixels to characterize them electrically and optically at low temperatures in a cryostat with optical access. In this chapter, we present the measurement setup including the cryostat, the optical setup and the readout electronics. We discuss the readout principle in order to determine the detector responsivity and show how the sensitivity can be calculated from the optical measurements. The performances are compared afterwards with the room temperature reflexion measurements and simulations.

8.1. The 132 pixel LEKID array

The LEKID geometry that showed the highest optical absorption properties, determined by room temperature reflection measurements and simulations, is the structure $LEKID3$ (see previous chapter). This geometry consists of a 3 μm wide meander line and a spacing between the parallel lines of $s = 276$ μm. Based on these results a 132 pixel LEKID array was designed and fabricated to measure it at cryogenic temperatures and to determine the sensitivity of the detectors. The resulting pixel design is shown in Fig. 8.1 presenting the single pixel design and a picture of a packaged 132 pixel array. The pixels have a size of 2.25×2.25 mm^2 including the ground plane frame. The distance between the pixels is therefore 2.25 mm (distance between pixel centers) also called the pixel pitch. The coupling to the transmission line is designed to $Q_C = 30000$ corresponding to $m = 16$ μm and $d = 16$ μm (see Fig. 6.2) guaranteeing a high loaded quality factor and sufficient dynamic range under variable background loading. The interdigital capacitor consists of 6 fingers of 10 μm width and 20 μm spacing. The frequency tuning is determined (in a linear range) to 1 $MHz/12$ μm finger length. The frequency spacing of the array is

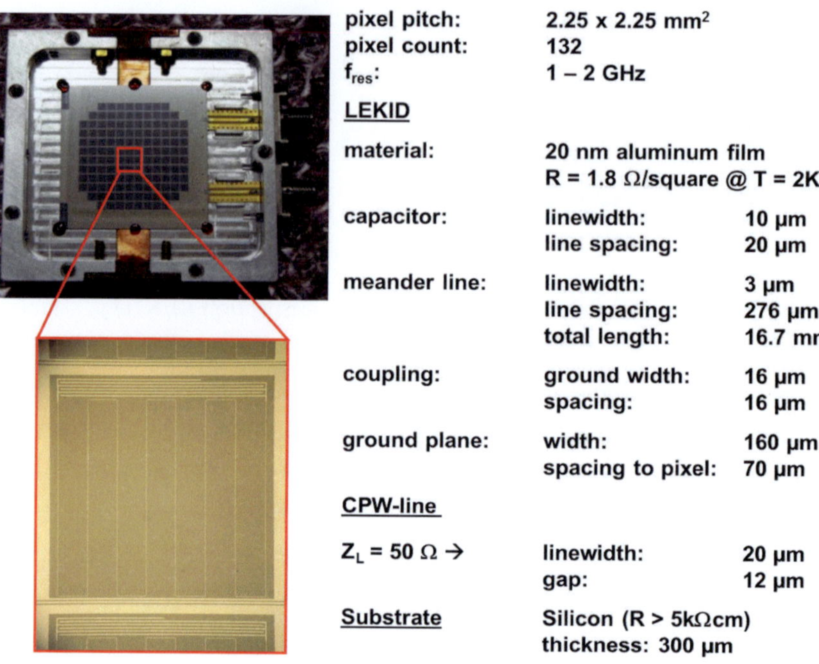

Fig. 8.1.: Pixel design based on the results of the electrical and optical optimization presented in the previous chapters. The shown LEKID array consists of 132 pixels. The array is mounted in a aluminum sample holder, where the ports are connected via bonding to 50 Ω cpw strips that are soldered to sma connectors.

between 1.5 and 2 MHz in order to pack the resonances densely in a 230MHz bandwidth but with sufficient space to avoid overlaps (double resonances). To reduce pixel crosstalk, a 160 μm wide ground plane between the pixels with 70 μm distance to the LEKID is chosen. For a ground plane of this width, the influence of the resonator distance is small and we can therefore use a relatively big distance (see Fig. 6.19) reducing the fabrication errors due to lithography problems. The cpw-line consists of a 20 μm center line with a 12 μm wide gap to achieve $Z_L = 50\ \Omega$. The transmission line is guided over the array as presented in Fig. 6.22, where the resonators are coupled from one side only to reduce cross coupling effects. The LEKIDs are made by a 20 nm aluminum film deposited on a 300 μm thick high resistivity (>5K Ωcm) silicon substrate. The sample holder in Fig. 8.1 is made of aluminum with two copper lines to achieve good thermalization of the LEKID array.

The aluminum becomes superconductive below 1.2 K and therefore a bad thermal conductor at sub-K temperatures. To connect the array to the external network, the transmission line is wire-bonded to 50 Ω strips that are soldered to sma connectors. The superconducting back-short is not shown in the picture but is integrated in the aluminum cover of the sample holder. The distance determined above is 700 μm for a center frequency of around 150 GHz (see room temperature reflection measurements of $LEKID3$).

8.2. Measurement setup

The setup to characterize these detectors optically at such low temperatures that are needed for high sensitivities of the KIDs ($T \approx T_C/10$), has been developed and built at the *Institute Néel* in Grenoble, France. This setup contains a dilution cryostat with optical access, the quasi-optical setup, the readout electronics and the data acquisition program. Here, we give a short but complete overview about this setup to explain how the LEKIDs have been characterized under almost real observation conditions including how to determine the sensitivity under different optical loads.

8.2.1. The NIKA cryostat

As discussed before, to obtain high sensitivity with KIDs, temperatures of $T << T_C$ are needed. To achieve these sensitivities in the case of aluminum we have to cool down the detectors to approximately $T = 100\ mK$, which demands an sophisticated cryogenic system. The cryostat not only has to cool down to such low temperatures, but it also has to keep the temperature continuously over a relatively long period of time. The NIKA cryostat works therefore on the basis of a ^{3}He/^{4}He-dilution refrigeration system. A mixture consisting of the two helium isotopes ^{3}He and ^{4}He is cooled down below 1 K, where the mixture separates by gravity into two liquid phases. One is the ^{3}He rich phase and the second one the ^{3}He poor phase with only 6 % of ^{3}He. If a ^{3}He atom of the poor phase crosses the boundary to the rich phase, this is only possible by taking energy from the environment leading to a decrease in temperature. Making this process continuously leads to a effective cooling down of the mixture. Usually the process ends when all the ^{3}He atoms of the rich phase crossed the boundary to the poor phase. To avoid this, we use a closed cycle system that allows to run the cryostat continuously stable at such low temperatures. ^{3}He

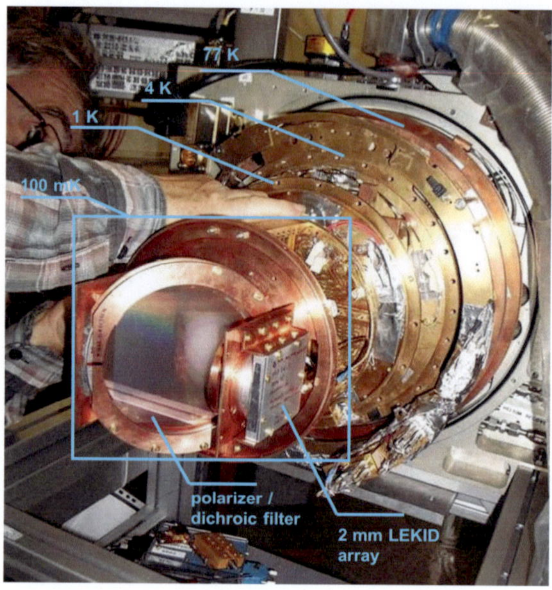

Fig. 8.2.: Picture of the open NIKA cryostat with view on the different temperature stages and the array mounting of the two frequency bands.

atoms that crossed the boundary and cooled the mixture are evaporated in a still afterwards. Outside the cryostat the gas is pumped to higher pressure and returns to the cryostat's condensator where the cycle restarts. A picture of the inner parts of the NIKA cryostat is shown in Fig. 8.2. The different temperature stages, thermally isolated from each other, are at 100 mK, 1 K, 4 K, 77 K, 150 K and 300 K. The 100 mK stage hosts a beam splitter that is either a polarizer or a dichroic filter, the lenses and the KID arrays for the two frequency bands. The cryogenic low noise amplifiers are mounted on the 4 K stage and exhibit an average noise temperature of around 4K. The cabling in the cryostat merely needs 4 coax cables to read out the two KID arrays and a few more wires for the thermometers and the amplifier power supply, which reduces the thermal connection between the different stages to a minimum.

8.2.2. The dual band optics

The direct detectors presented here detect power over a relatively large frequency band. We must therefore define this bandwidth by placing proper filters in the beam path. The optics in the cryostat including lenses and filters are adapted and optimi-

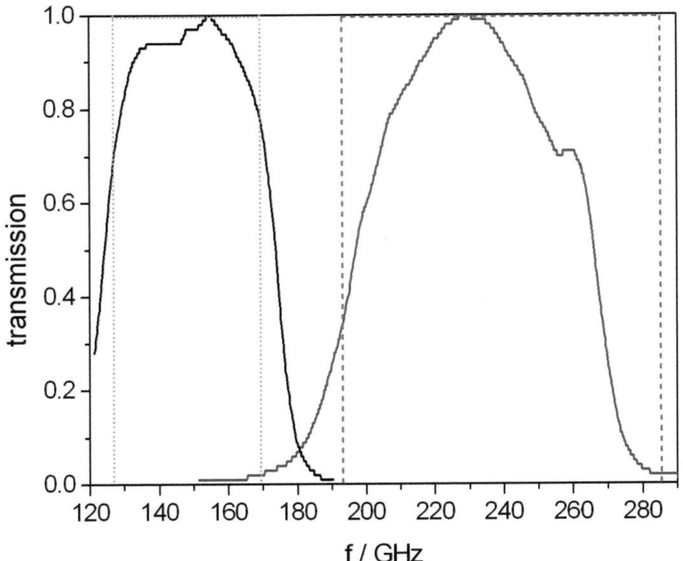

Fig. 8.3.: Normalized total mm-wave transmission of the NIKA filter setup for the 1.25 and the 2.05 mm band. The filter transmission is compared to the 3 dB bandwidth of the atmospheric opacity at Pico Veleta.

zed to the beam path of the IRAM 30 m telescope. The filters used in the NIKA cryostat have been fabricated by the $Astronomy\ Instrumentation\ Group\ (AIG)$ at $Cardiff\ University$ to provide the two frequency bands and to limit the optical power that enters the cryostat in order to avoid heating it up. The most selective filters are therefore mounted at the coldest stage to guarantee that radiation from hotter sources does not reach the detectors. The resulting transmission at the two frequency bands at 1.25 and 2.05 mm have been measured using a Martin Puplett interferometer and is shown in Fig. 8.3. The optical transmission (normalized) defined by the filters is compared to the -3 dB bandwidth of the corresponding atmospheric windows at Pico Veleta. We can see that the filter at the 2.05 mm band is well defined and does not limit the available bandwidth defined by the atmosphere, whereas at the 1.25 mm band we lose around 10 GHz of bandwidth at the highest frequencies. This gives us an effective bandwidth for the two bands of $\Delta f_{1.25\ mm} = 90\ Ghz$and $\Delta f_{2.05\ mm} = 40\ GHz$.

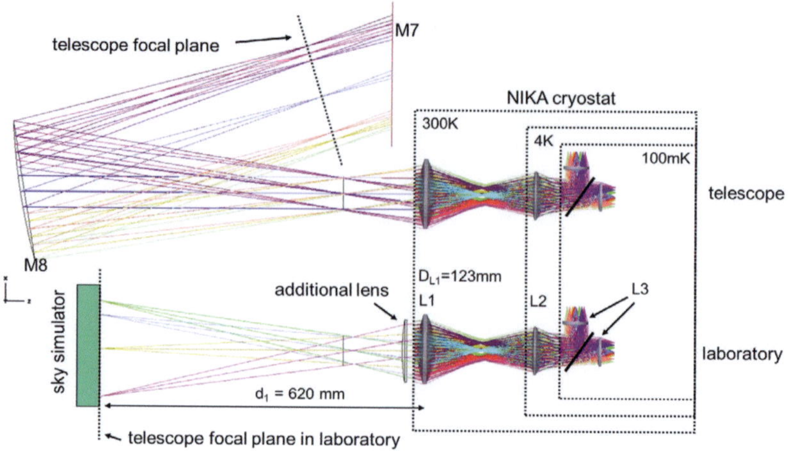

Fig. 8.4.: ZEMAX simulations of the optical beam path in the telescope cabin and the laboratory. Three lenses (L1,L2,L3) are used to focus the beam to the detector plane. d_1 is the distance from L1 (with D_{L1}=123 mm) to the laboratory focal plane defining the solid angle of the system.

In Fig. 8.4 simulations of the beam path at the telescope site and in the laboratory are presented. These simulations have been done by Samuel Leclercq using ZEMAX [1] including all lenses, filters and the beam splitter. There are 3 corrugated lenses used to focus the beam, the first at the 300 K stage (L1), the second at the 4 K stage (L2) and the third one at the 100 mK stage (L3) just in front of each detector array. On the upper side, the figure shows the beam path starting at the last two mirrors (M7 and M8) in the telescope cabin before entering the cryostat and indicates the telescope focal plane. On the lower side, a so-called sky simulator with calculated diameter and distance is placed in front of the cryostat. To obtain the same focal plane as at the telescope but closer to the cryostat window, an additional lens is mounted outside the cryostat allowing to characterize the detectors in the laboratory under almost real observing conditions. The sky simulator is explained in more detail below.

Fig. 8.5 shows a zoom on the 100mK stage including the sample holders of the two arrays at 2.05 mm (a) and 1.25 mm (b), where just in front of the arrays the third lens and the most selective filter is mounted. For the measurements in this chapter, a grid polarizer (c) was used as beam splitter, which reduces the power on each pixel by a factor of 2 and has to be taken into account for the determination

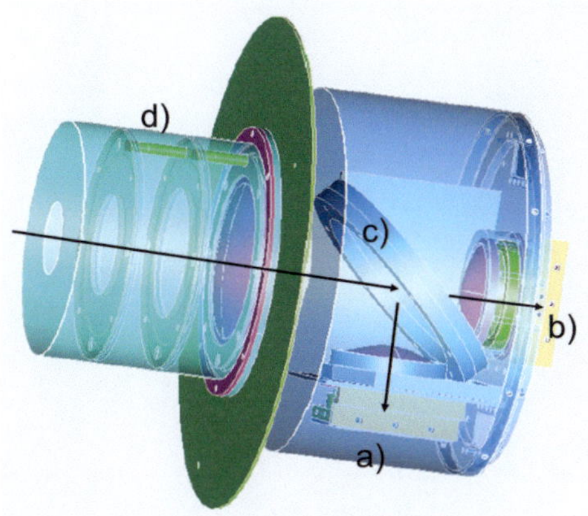

Fig. 8.5.: Schematic of the 100 mK stage optics of the NIKA cryostat. a) detector array for the 2 mm band; b) detector array for the 1mm band; c) band splitter (polarizer or dichroic filter); d) baffle

of the detector sensitivity. The polarizer is a simple and possible solution since the detectors of both bands are sensitive to only one polarization. To reduce off-axes radiation hitting the detectors, a multi-stage, black coated baffle (d) is installed in front of the 100 mK stage. The total transmission t_{total} of the optics in the cryostat without taking the polarizer into account is

$$t_{total} = t_{filters} \cdot t_{lenses} \approx 40\% \qquad (8.1)$$

Including the polarizer we have an effective transmission of around 20 % of power that is hitting each detector array.

8.2.3. The sky simulator

The sky simulator mentioned above is used to simulate the sky background at the telescope site. A picture of it is shown in Fig. 8.6 on the left hand side. It consists of a 24 cm black body disk whose temperature can be varied from 50 to 300 K. The disk is mounted inside a pulse tube cryostat with minimum temperature of 50 K. Once the minimum temperature is reached, it can be varied by a heating resistor. This allows simulating the optical load of the LEKIDs under different sky conditions, which corresponds to around 50 K under normal weather conditions. The cryostat

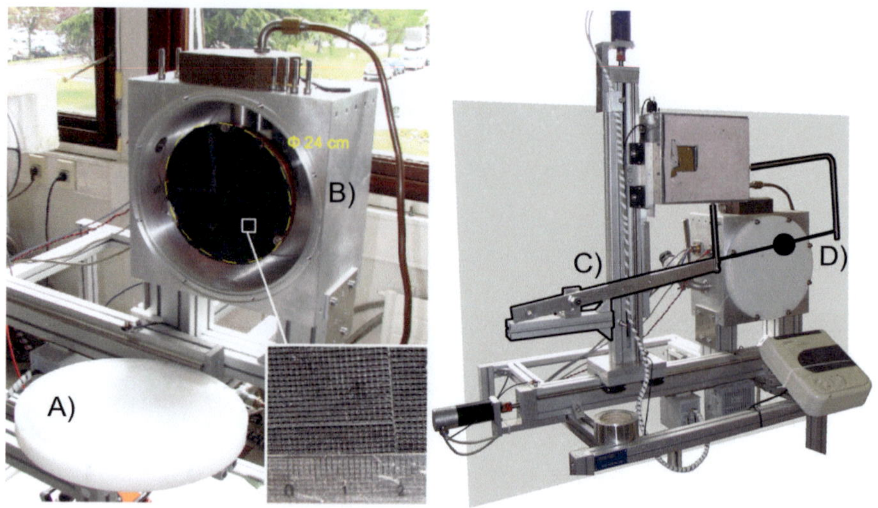

Fig. 8.6.: Picture of the sky simulator setup: A) HDPE window; B) pulse tube refrigerator with black body disk; C) In XY-direction movable arm; D) artificial planet (high-ε material).

is closed with a HDPE window (A), which has very little optical loss. To simulate the observation of a real source, a small ball (D) is mounted on a XY-table (C) as shown in Fig. 8.6 on the right hand side. The size of this artificial planet correspond to the size of Mars on the focal plane, which is about 8 mm of diameter. This, in X and Y direction movable artificial planet with black body temperature of around 300 K is used to simulate the observation of a real astronomical source. The background is hereby defined by the temperature of the sky simulator and the source by the size and temperature of the small planet. This setup is used to determine the sensitivity and position of each pixel individually. By moving the small planet in front of the cryostat, the maximum detector response is found when the artificial planet crosses the center position of each pixel. This way, the pixel position on the focal plane can be determined for each pixel. More details about this setup can be found in [8].

8.2.4. The readout electronics

One of the advantages of the KID technology is the capability to read out simultaneously a large number of pixels over a single transmission line via *frequency multiplexing*, allowing to build large filled array cameras for radioastronomy. The

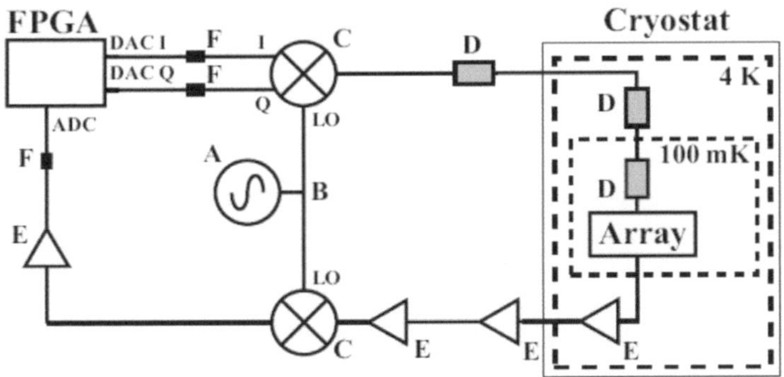

Fig. 8.7.: Schematic of the readout electronics for the NIKA 2.05 mm band taken from [59]. A) local oscillator (LO); B) splitter; C) IQ-mixer; D) attenuator; E) amplifier; F) low pass filter. Not shown is the computer connected to the FPGA.

multiplexing of other existing bolometer arrays such as MAMBO[1] or BOLOCAM[2] demands a individual readout circuit for each pixel, which leads to a large number of cables entering the cryostat. In addition the fabrication of this, usually time domain multiplexed readout circuit using SQUIDs (Superconducting QUantum Interference Detector), is much more complicated and expensive. The scaling is hereby limited by the multiplexing factor of the readout electronics. The KIDs offer the possibility to multiplex the detectors in frequency, which makes the number of pixels that can be read out dependent on the available frequency bandwidth, limited by the analog to digital converter (ADC) sampling rate, and pixel spacing. In this section, we present the NIKA readout electronics and how the digital processing is realized.
Fig. 8.7 shows a schematic of the entire resonator excitation and readout loop of the NIKA KIDs. The tone of each resonance frequency is generated in a *Field Programmable Gate Array* (*FPGA*) in a bandwidth of 230 MHz. This FPGA board including a 16 bit digital to analog converter (DAC) and a 12 bit analog to digital converter (ADC) with 500 MSPS was developed in the framework of a collaboration initialized by the *University of California Santa Barbara* (*UCSB*). These electronics (ROACH board) developed for KIDs and the data acquisition program developed at the *Institut Néel* allow the simultaneous readout of 112 pixels in a bandwidth of 230 MHz. The tones are hereby generated as I (inphase) and Q (qua-

[1]Instrument based on semi-conductor bolometers for 1 mm at the IRAM 30 m telescope [47]
[2]Bolometric camera for mm-wave observations at CSO [25]

drature) component with a phase difference of 90° between both. The resonance frequencies of the LEKIDs are between 1 and 2 GHz depending on the geometry and material parameters, which means that the generated tones have to be mixed up to this frequency range. This is done using an IQ-mixer (C) and a local oscillator (LO) signal (B). The power of the input signal after the IQ-mixing is limited by an USB controllable attenuator (D) in order to avoid a saturation of the LEKIDs. Typical input powers that have been used, depending on the geometry and the aluminum film thickness of the detectors, are around $P_{input} \approx -75\ dBm$. At the output of the LEKID array, the signal is amplified by a cryogenic low noise amplifier (D) at the 4 K stage and further amplified at room temperature (E). This output signal is then mixed down to the original FPGA bandwidth using a down converter and the same LO signal. The power of this signal is then adapted to the dynamic range of the following ADC using another controllable attenuator.
The principle of the frequency up and down mixing is illustrated in Fig. 8.8 for two resonance frequencies. To generate the frequency comb in the FPGA using a *Cordic* algorithm, the resonance frequency of each resonator has to be known. Therefore, we measure the transmission scattering parameter S_{21} using a network analyzer. Taking this data, the frequency comb is generated in the mentioned bandwidth of 230 MHz respecting the measured frequency spacing. To demonstrate the principle, only two tones (ω_1 and ω_2) are generated. The frequency of the LO (ω_{LO}) is also determined from the S_{21} measurement and is located a couple of MHz below the lowest resonance frequency. Using then an IQ-mixer, we obtain the new tones as shown in the schematic at $\omega_1 + \omega_{LO}$ and $\omega_2 + \omega_{LO}$. The output time domain signal, modified in amplitude depending on the detected signal, is then mixed down using a second IQ-mixer to the FPGA bandwidth before it enters the ADC of the FPGA board. The digital signal from the ADC is then split in 2 branches and multiplied by the sine and cosine waves of the tone of interest (an image of the input signal before the DA conversion). This way, the inphase (I) and quadrature (Q) component of the tone are obtained and thus the amplitude (A) and phase ϕ of each transmitted tone given by

$$A = \sqrt{I^2 + Q^2} \quad and \quad \phi = arctan\left(\frac{Q}{I}\right) \tag{8.2}$$

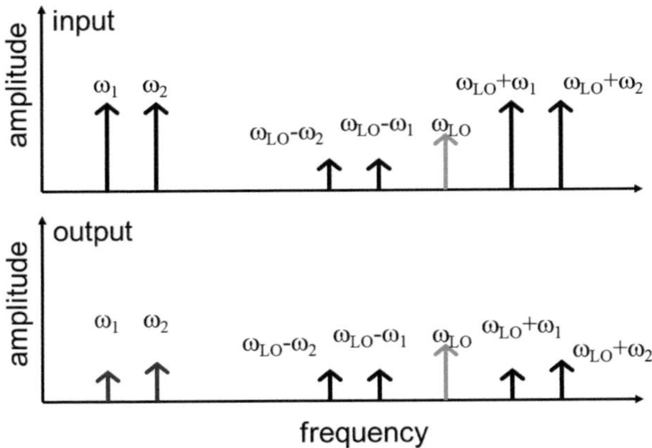

Fig. 8.8.: Principle of frequency up and down mixing using an IQ-mixer and a LO signal for two resonance tones.

can be monitored continuously. This readout principle is described in more detail in [10].

Another approach is to plot these values in the complex IQ-plane. An example of a frequency sweep around a resonance for two different optical loads is shown in Fig. 8.9. The black curve represents the resonance curve with the lowest optical load, where no signal is detected. The value at the resonance frequency in the IQ-plane is given by R. For an illumination of the detector, the radius of the circle decreases slightly and the resonance point moves around the circle to the new point D. As we see, the change in radius is much smaller and can be neglected for small signals allowing to define a relative change in phase corresponding to the angle $\Delta\phi$ about the center of the curvature C. From the value of the circle center, determined by a procedure implemented in the data acquisition program, and the measured I_D and Q_D values we can calculate this angle to

$$\Delta\phi = arctan\left(\frac{Q_D - Q_C}{I_D - I_C}\right) - \phi_0 \tag{8.3}$$

where (I_C, Q_C) defines the center of the resonance circle C and (I_D, Q_D) the measured values due to illumination of the detector. ϕ_0 rotates the plane such that the curve intersects (at point R) the Q-axis at the resonance frequency f_{res}. The measu-

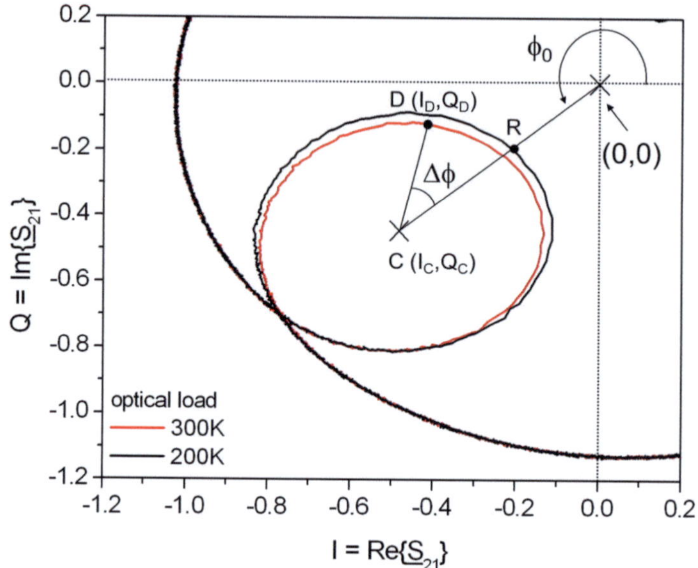

Fig. 8.9.: Readout principle of amplitude and phase plotted in the IQ-plane. The resonance circle with center C and resonance frequency point R is modified when the detector is illuminated leading to a new resonance point D. The resulting angle $\Delta\phi$ is called the relative phase that can be calculated from the measured I and Q.

rement results presented later on have been obtained using this readout method by determining the LEKID response in relative change in phase $\Delta\phi$.

The response of every pixel of the array is measured simultaneously and broadcast via UDP packets by the ROACH electronics to the control computers at a rate of 22 Hz, where the data processing is done. This program is called CAMADIA and was being developed at the *Institut Néel* as well but will not be described here. For more information about it, see [8].

8.3. Photometry and sensitivity calculations

8.3.1. Photometry

Photometry is a technique of astronomy measuring the flux or the intensity of an astronomical source's electromagnetic radiation. In the case of ground based observations, it describes how much radiation is gathered by the telescope, passing the lenses and filters in the camera and is finally hitting the detector array. To determine the sensitivity of the one LEKID, we have to calculate this radiation that illuminates the pixels, which depends on different parameters such as the emissivity of the astronomical source defined by its size and temperature (black body radiation), the optics of telescope and camera and the frequency band in which the observations are done (1.25 and 2.05 mm band for NIKA). We give here a short overview of the photometry calculations to estimate the fraction of power of a black body source hitting one pixel. For more precise values of the optical calculations we used the simulation tool *ZEMAX*.

Black body radiation The calculation of the black body radiation of an astronomical object allows the determination of the photometry for the NIKA instrument. For resolved sources, which means objects that are already well studied and its flux is therefore known, it also gives the possibility to calibrate the NIKA instrument. The power density emitted by a perfect black body source (emissivity $\varepsilon = 1$) depending on its surface S, temperature T, the frequency band $d\nu$ with center at ν and under a solid angle Ω with an emission angle β is given by *Planck's law*:

$$P(T,\nu,\beta) = \int\int\int \frac{2h\nu^3}{c^2} \frac{1}{e^{(\frac{h\nu}{k_BT}-1)}} cos(\beta) dS d\nu d\Omega \quad [Wm^{-2}Hz^{-1}sr^{-1}] \qquad (8.4)$$

This expression is not integrable in an analytical form but can be approximated under certain conditions. Here we use the so-called *Rayleigh-Jeans approximation* describing the radiation of a black body dependent of the frequency ν at a given temperature T. This approximation giving a linear relation between power and temperature is valid for $k_BT >> h\nu$ and allows therefore the determination of the sensitivity in terms of temperature, which is often preferred for astronomical sources and easier to derive. Inserting this approximation in 8.4 we obtain a simplified expression for the power density:

$$P(\nu,\beta) = \int\int\int \frac{2k_B\nu^2}{c^2} T cos(\beta) dS d\nu d\Omega \quad [Wm^{-2}Hz^{-1}sr^{-1}] \tag{8.5}$$

Taking the surface of the black body into account, given by $S = \pi D_s^2/4$ with its diameter D_s, we can write the power density as

$$\Delta P(\nu,\beta) = \int\int \frac{2k_B\nu^2}{c^2} \frac{\pi D_s^2}{4} \Delta T cos(\beta) d\nu d\Omega \quad [WHz^{-1}sr^{-1}] \tag{8.6}$$

We define here $\Delta T = T_s - T_{bg}$ as the temperature difference between the source (T_s) and typically the background temperature (T_{bg}).
Only a fraction of the total emitted power reaches the detector plane depending on the focusing and the effective transmission of the optics (filters and lenses). The opening angle is defined by the geometrical dimensions of the optics and the distance to the source. This defines the solid angle Ω of our aperture and modifies the power density to

$$\Delta P(\nu) = \int \frac{2k_B\nu^2}{c^2} \frac{\pi D_s^2}{4} \Delta T \frac{\Omega}{2} d\nu \quad [WHz^{-1}] \tag{8.7}$$

where the solid angle is given by $\Omega = 2\pi(1 - cos(\psi))$ with the apex angle $\psi = arctan(D_L/2d_1)$. D_L is in our case the diameter of the lens L1 at the 300K stage and d_1 is the distance to the focal plane (see Fig. 8.4). For our laboratory setup we obtain $D_L = 123\ mm$ and $d_1 = 620\ mm$ giving $\psi \approx 5.6\ deg$.
As mentioned above, the filters and lenses in the cryostat reduce the optical transmission depending on the wavelength of the incoming signal. Taking this total transmission $t_{total} \approx 40\ \%$ (see above) into account and integrating the power density over the bandwidth of the 2.05 mm band we obtain

$$\Delta P = \frac{2k_B}{c^2} \frac{\pi D_s^2}{4} \Delta T \frac{\Omega}{2} \frac{\nu_{max}^3 - \nu_{min}^3}{3} t_{total} \quad [W] \tag{8.8}$$

where $\nu_{max} = 170\ GHz$ and $\nu_{min} = 125\ GHz$. This corresponds to the power emitted from the surface of the sky simulator reaching the detector focal plane.

Point source If we consider a source, whose angular size is much smaller than the angular beam size defined by its *Full Width Half Maximum* (*FWHM*) we can speak of a point source (or unresolved source). The *FWHM* defines the beam diameter where the maximum power is reduced by a factor of two. The diffraction pattern of such point sources can be described by a *Point Spread function* (*PSF*), which can be approximated using a 2D Gaussian normal distribution given by

$$PSF(r) = e^{-\frac{r^2}{2\sigma}} \quad with \quad \sigma = \frac{FWHM}{2\sqrt{2ln2}} \tag{8.9}$$

For the calculation of the power fraction on a single pixel we have to consider two different cases: a) the temperature dilution due to the diffraction of the point source in the plane of the source and b) the PSF oversampling in the detector plane. These two cases are presented in Fig. 8.10 for the example of Mars in October 2010 with angular diameter $D_s = 4''$ and surface temperature of T=200 K. On the left hand side the principle of temperature dilution is demonstrated. The size of the instrumental beam is given by a $FWHM = 16''$ for the NIKA optics at the 2.05 mm band. Since

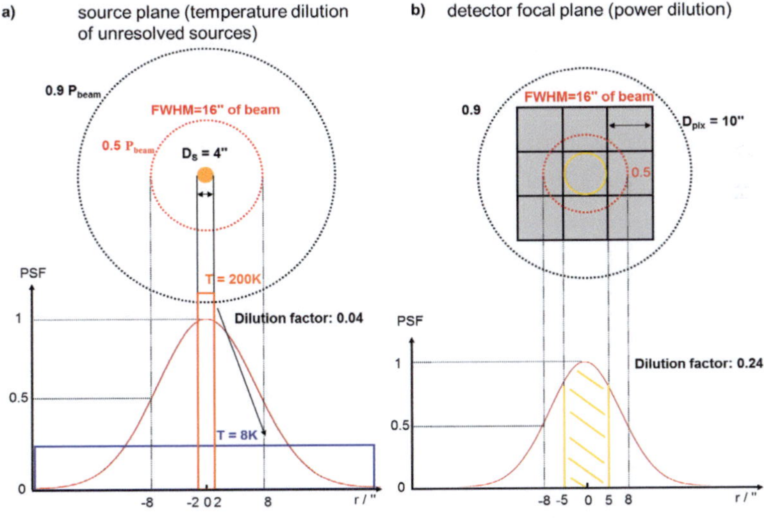

Fig. 8.10.: Dilution of the source temperature (a) and power (b) due to the diffraction pattern of the source and the beam size that is larger than the source and pixel. The diffraction can be described in both cases by the PSF approximated by a 2D Gaussian normal distribution.

this is larger than the size of the source, the emitted power is spread over the entire beam size, which decreases the overall temperature to a smaller value following the conservation of energy. As mentioned above, this diffraction is described by the PSF that is also shown in the schematic. Calculating now the effective temperature over the entire beam size that corresponds to the source temperature (200 K , we obtain a so-called temperature dilution factor. This factor can be determined by integrating the PSF over the surface of the source and normalizing the expression, whose integral is given by

$$\Gamma(r) = \frac{1}{2\pi\sigma^2}\int_0^{2\pi}\int_0^{x} PSF(r) r dr d\theta = 1 - e^{-\frac{r^2}{2\sigma}} \tag{8.10}$$

For the point source example in Fig. 8.10 with $r = D_s/2 = 2''$ and $FWHM = 16''$ we calculate a dilution factor of around 0.04, which means that the effective temperature of the beam is $T = 200\ K \cdot 0.04 = 8\ K$. The principle of this temperature dilution is shown in the schematic by the two rectangles representing the power per surface. The source has a temperature of T=200 K over a width of its diameter $D_s = 4''$ (orange). The effective temperature of the beam corresponds then to a height of a new rectangle (blue) covering the entire beam size.

Since this beam with new effective temperature is larger than the size of one pixel on the detector array, the power is spread over more than one pixel. The effective power illuminating one pixel is therefore again reduced and has to be taken into account. The Gaussian beam power distribution on the detector plane follows hereby, as a good approximation, the same normal distribution used for the source. This is demonstrated in Fig. 8.10 on the right hand side. The beam size is again defined by the $FWHM$ and we can see that it overlaps more than one pixel with diameter $D_{pix} = 10''$ or 2.25 mm. To calculate the fraction of power that is illuminating one single pixel, we integrate the PSF (using 8.10) over the size of one pixel ($r = D_{pix}/2 = 5''$), assuming a circular pixel surface (see yellow surface under PSF curve). The result of this integration corresponds to the power dilution factor due to a pixel size that is smaller than the beam size. For the example in the schematic we calculate a factor of around 0.24, which means that the beam power is distributed over around 4 pixels.

The total power on one pixel, taking the two dilution factors into account, is then given by

$$\Delta P_{pixel} = \Delta P \cdot \Gamma_{source}(r) \cdot \Gamma_{pixel}(r) \qquad (8.11)$$

In general the temperature ΔT before the dilution is unknown except in the case of planets that are already observed at other wavelengths (resolved sources). The sky simulator and the artificial planet temperature are well known and we can calculate the power on a single pixel for the laboratory setup.

8.3.2. Pixel sensitivity

We determine the sensitivity of the NIKA LEKIDs in *noise equivalent power* (NEP), which is defined as

$$NEP_{optical} \doteq \frac{\Delta P_{pixel}}{SNR} \quad [W/\sqrt{Hz}] \qquad (8.12)$$

where ΔP_{pixel} is the power obtained as shown above. The NEP is defined as the input power of a system required to obtain a signal to noise ratio equal to 1 in a 1 Hz bandwidth. This means that only signals with power that corresponds to multiples of the NEP can be measured ($P_{optical} = n \cdot NEP$ with $n = 1, 2...$). We call this sensitivity $NEP_{optical}$ since it is not a pure electrical NEP characterizing the detectors, but takes also the photometry of the system into account. The *signal to noise ratio* (SNR) is the detector response ΔS to the power ΔP_{pixel} divided by the spectral noise density $S_n(f)$ at a given frequency f:

$$SNR = \frac{\Delta S}{S_n(f)} \quad [\sqrt{Hz}] \qquad (8.13)$$

It is often easier, especially in the case of the sky simulator and its known temperature, to determine the temperature variation ΔT_{pixel} allowing to calculate a temperature dependent sensitivity given by the *noise equivalent temperature* (NET). This sensitivity is defined in analogy to the NEP with

$$NET_{optical} = \frac{\Delta T_{pixel}}{SNR} \quad [K/\sqrt{Hz}] \tag{8.14}$$

The presented sky simulator allows to measure the NET relatively easy. If we take the spectral noise density at a fixed background temperature defined by the black body disk and increase then the temperature by ΔT to determine the detector response ΔS, the NET can be calculated with 8.13. ZEMAX simulations have shown, that the optical loading P_{pixel} per Kelvin (for the NIKA-sky simulator configuration for the 2.05 mm band) is linear with $P_{pixel}/10\ K \approx 1 \cdot 10^{-13}\ W/K$. To simulate a typical background condition at the telescope site of $T \approx 50\ K$, we cool down the sky simulator to this temperature giving an approximated optical pixel load of around $P_{pixel} \approx 5\ pW$. Taking into account the additional power due to stray light (off axis 300 K radiation) coming from the laboratory environment, the total power on one pixel is around $P_{pixel} = 9\ pW$. This corresponds to an additional temperature of 40 K giving a total background temperature per pixel of 90 K, which has to be taken into account for calculating the photon noise of the LEKIDs.
The NET and NEP are therefore linked by the ratio of the power on one pixel and the corresponding *Rayleigh-Jeans* black body temperature difference ΔT_{RJ}:

$$NET = NEP \frac{\Delta T_{RJ}}{\Delta P_{pixel}} \tag{8.15}$$

In the case of the NIKA laboratory setup at 2.05 mm, we calculate a conversion factor of $\Delta T_{pixel}/\Delta P_{RJ} \approx 10^{13}\ K/W$.
With these photometric calculations and the laboratory setup we are now able to determine the sensitivity in NEP and NET of each individual LEKID on the 132 pixel array for different optical sources.

8.4. Low temperature measurements of the 132 pixel LEKID array

To optically characterize the LEKIDs at low temperatures, the 132 pixel array has been mounted into the NIKA cryostat and has been cooled down to T=100 mK. We present and discuss in this section the characterization of the LEKIDs including the electrical properties of the resonators, the optical efficiency at such low temperatures and the LEKID sensitivity.

8.4.1. Resonator properties at T=100 mK

Before determining of the detector sensitivities, we performed a measurement of the S_{21} scattering parameter using a network analyzer. Fig. 8.11 shows the scan of S_{21} that has been done with an open cryostat looking at a black body temperature of the sky simulator of T=50 K. This measurement is performed for two reasons: To determine the resonance frequency comb for the readout electronics and to check the quality factors and shape of the resonances. As mentioned before, the power on the array input is at $P_{in} \approx -75\ dBm$ and has a non negligible influence on the shape of the resonances. A too high input power can distort the resonance shape due to saturation effects and has therefore to be well adapted. We measured here

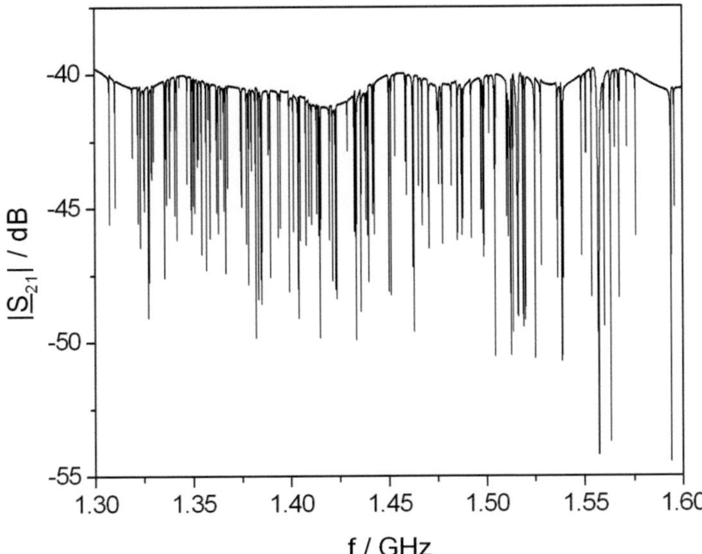

Fig. 8.11.: Measurement of the S_{21} of the 132 pixel LEKID array at T=100 mK. The measurement is done with an open optical access making the detectors looking at a 50 K black body temperature of the sky simulator plus 40 K coming from stray light corresponding to a single pixel load of $\Delta P_{pixel} \approx 9\ pW$.

an average intrinsic quality factor of $Q_0 \approx 150000$ and a coupling quality factor of $Q_C \approx 25000$, which corresponds perfectly to the designed values. From the 132 pixels on the array, 123 resonances have been identified. This means that around 5 % of the resonators do not work, which has mostly fabrication reasons (lithography or etching). The resonances are within a bandwidth of around 290 MHz, exceeding the 230 MHz available from the readout electronics. This leads to a loss of around 15 resonances that cannot be used for observations. The problems due to pixel cross coupling are higher than expected but acceptable which can be seen from the S_{21}-scan. The resonances are not homogeneously spaced, but the number of double resonances is relatively small. We calculate an average pixel spacing of 2.2 MHz that is slightly higher than the designed 1.5-2 MHz leading to an exceeding of the electronics bandwidth. The variations of the off resonance transmission at around -40 dB is caused by the length of the transmission line and reflections from the cold amplifier and the bondings due to impedance mismatches.

8.4.2. Optical efficiency at T=100 mK

The optical efficiency has been determined using simulations and room temperature reflection measurements presented in Chapter 7. At the *Institute Néel* we have the possibility to perform *Fourier Transform Spectroscopy* (*FTS*) measurements using a Martin Puplett interferometer adapted to the NIKA cryostat that has been done in the framework of a former PhD thesis [18]. In Fig. 8.12 a normalized FTS measurement of the average pixel response depending on the frequency in the 2.05 mm band is shown. This measurement (dashed blue) is compared to the obtained results from the simulation (dotted red) and reflection measurement (black solid). The FTS results confirm the applicability of the CST simulations and room temperature measurement by excellent agreement between both. The simulated and measured optical efficiency is compared to the -3 dB bandwidth of the NIKA filters (green line) where we can see that the filters are limiting the effective bandwidth below 125 GHz and above 175 GHz. The FTS curve is normalized because the measurement results are not in absolute values due to the unknown optical load in the NIKA FTS configuration. It is therefore difficult to compare the absolute absorptivity. The effective bandwidth on the other hand is well comparable and shows good agreement.

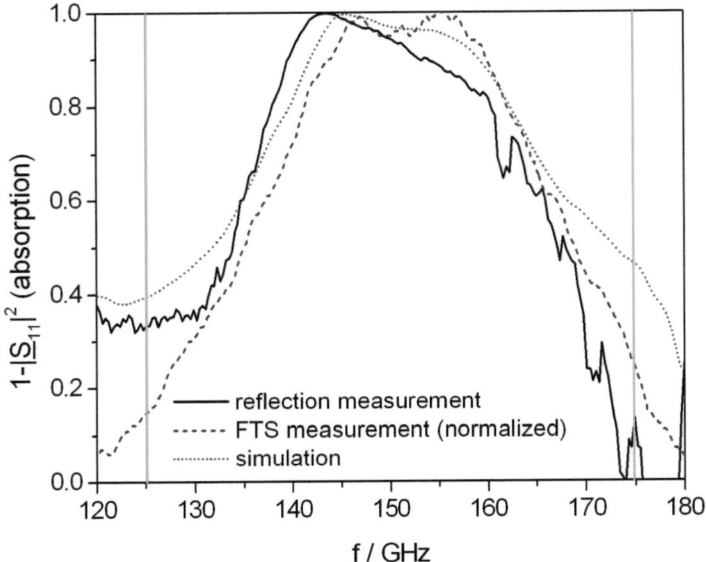

Fig. 8.12.: Room temperature measurement (black solid) and simulation of the optical efficiency (red dotted) compared to a FTS measurement (blue dashed) done with an Martin Puplett interferometer. In green: the -3 dB bandwidth of the NIKA filters for the 2.05 mm band.

8.4.3. Sensitivity of the NIKA LEKIDs

To determine the LEKID sensitivity, we calculated first the optical NET as described above. For this, we measured the spectral noise density of each pixel over the bandwidth from 0.1 to 10 Hz. The noise spectrum is obtained by performing a Fourier transformation of the detector time domain phase signal corresponding to the frequency noise. Such a spectrum for the 132 pixel array is shown in Fig. 8.13 and is given in $Hz/\sqrt{Hz}$. We can distinguish between two principal noise sources, the detector noise and the sky noise, caused by temperature fluctuations of the background (in our case the sky simulator black body). The low frequency sky noise (green line) has a 1/f-behavior and dominates the total noise up to 0.4 Hz. The detector noise (red line) on the other hand is relatively flat between 0.4 and 10 Hz. The measured noise spectrum is compared to a theoretical estimated value for the photon noise which defines the lower limit of the detector noise (see Chapter 3). The average

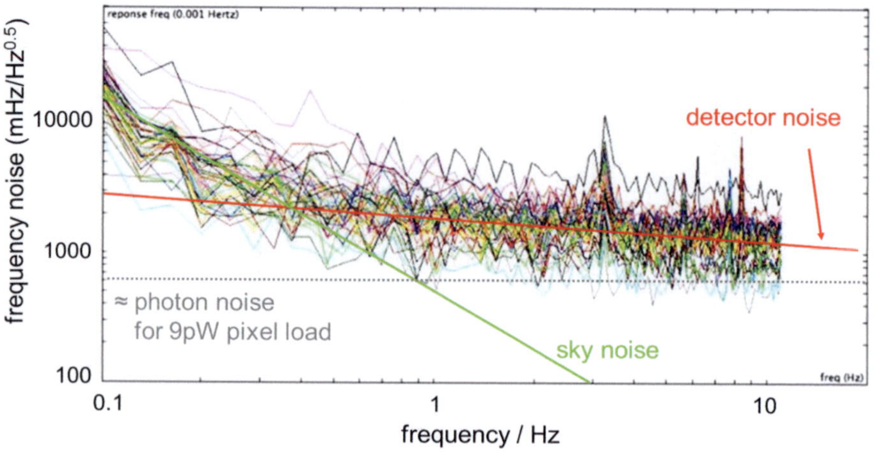

Fig. 8.13.: Frequency (phase) noise spectrum of each pixel calculated from the time domain data by a Fourier transformation between 0.1 and 10 Hz. Red line: detector noise; green line: sky noise; dashed gray line: photon noise limit for equivalent optical load.

pixel noise at a standard representative frequency of 1Hz is $S_n(1Hz) = 2\ Hz/\sqrt{Hz}$. Estimating the photon noise for an equivalent pixel load to $S_{n,photon} \approx 0.6\ Hz/\sqrt{Hz}$, we find that the measured noise is around a factor of 3 higher than this limit. A noise that close to the fundamental limit is already a promising result regarding the sensitivity of the LEKIDs. Comparing the LEKID noise spectrum at 10 Hz, we even see some pixels that are already limited by the photon noise. The measured noise spectrum and the noise limit take into account the polarizer used to split the incoming signal for the two detector arrays, which reduces the optical power hitting the pixel again by a factor of two. Taking the numbers given above, this corresponds to a $\Delta P_{pixel} \approx 4.5\ pW$ and T = 90 K background temperature (sky simulator + stray light).

To calculate the sensitivity of the LEKIDs using the measured noise spectrum we must determine the LEKID responsivity. For this, the sky simulator temperature has been increased by $\Delta T = 5\ K$ and the corresponding detector response (frequency or phase shift) was measured. Hereby, an average detector response of $\Delta S = 2.2\ kHz$ (445 Hz/K) was found leading to a signal to noise ratio of $SNR = 1100\ \sqrt{Hz}$ and an average noise equivalent temperature of $NET_{optical} = 4.5\ mK/\sqrt{Hz}$ at 1 Hz. In Fig. 8.14 the NET of each individual measurable LEKID is shown alongside the average value of all pixels (dashed blue line). It demonstrates that the individual sensitivity, regarding the first 60 pixels, varies between 2 and 10 $mK/\sqrt{Hz}$. The higher NET of the last 30 pixels can be explained by input power fluctuations caused in the electronics during the generation of the resonance frequency comb. The last group of resonators are exited with a power that is roughly 3-4 dB lower, leading to an increased noise level.

We can now calculate the corresponding sensitivity in the form of noise equivalent power (NEP) using the conversion factor presented above. For $\Delta T = 5\ K$ of the sky simulator we determine a corresponding power $\Delta P_{pixel} = 0.5\ pW$ leading to $NEP_{optical} = 5 \cdot 10^{-16}\ W/\sqrt{Hz}$. Since the polarizer has no influence on the temperature that the pixel sees, but on the power per pixel, we can divide the calculated NEP by a factor of 2 giving an average $NEP_{optical} = 2.5 \cdot 10^{-16}\ W/\sqrt{Hz}$. It is worth to mention that the best LEKIDs of the 132 pixel array showed sensitivities in the range of $NEP_{optical} = 7.5 \cdot 10^{-17}\ W/\sqrt{Hz}$ corresponding to $NET_{optical} = 1.5\ mK/\sqrt{Hz}$. These numbers are competitive with today's existing bolometer instruments working in the same frequency range.

Based on these excellent results, IRAM decided to test the NIKA dual band camera with the presented 2.05 mm LEKID array and a 1.25 mm SRON array based on antenna coupled MKIDs during a second test run at the 30 m telescope.

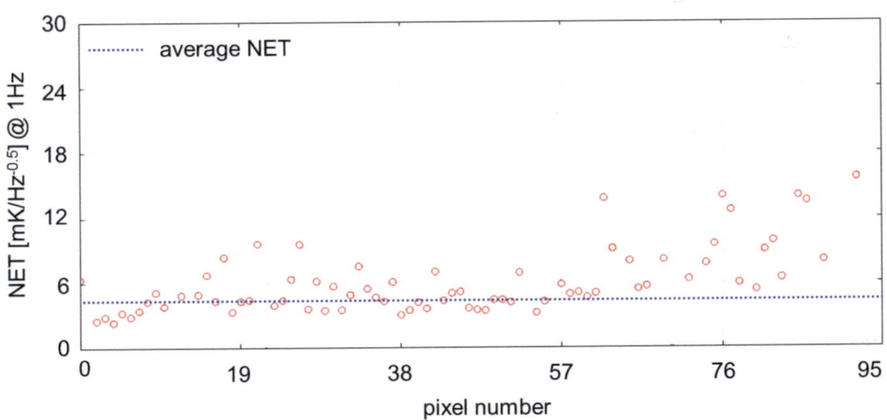

Fig. 8.14.: Noise equivalent temperature (NET) of each measurable LEKID of the 132 pixel array. The average NET taken from the first 80 pixels is around 4.5 $mK/\sqrt{Hz}$ at 1Hz.

9. NIKA LEKIDs at the IRAM 30 m telescope

A first test run at the telescope was performed in 2009 with a single band optics allowing the measurement of 30 LEKID pixels [45]. Here we present the results of the second NIKA telescope engineering run that took place in October 2010 for the first time allowing to do observations with two bands simultaneously [44]. For the 2.05 mm band, a 144 pixel LEKID array was mounted whereas at the 1.25 mm band a KID array developed at SRON based on antenna coupled MKIDs was used. We concentrate here on the results of the LEKID array whose development is discussed throughout this thesis.

9.1. Installation of NIKA at the telescope

The NIKA cryostat is engineered to fit the receiver cabin of the IRAM telescope in Pico Veleta, Spain. The 30-meter primary mirror (M1) and the hyperbolic secondary (M2) (see Fig. 2.1) are installed directly on a large alt-azimuth mounting and the incident beam is thus directed into the receiver cabin through a hole in M1 using a standard Cassegrain configuration. A rotating tertiary (M3) provides a fixed focal plane (Nasmyth focus). The optical axis, in order to conform to the dimensions of the cabin, is deviated by two flat mirrors (M4 and M5). NIKA re-images the large telescope focal plane onto the small sensitive area covered by the KIDs as explained in 8.2.2. The demagnification factor is around 6.6, achieving a well-adapted scale of 5 arcsec/mm on the detector plane. This is accomplished using two flat mirrors (M6, M7), one bi-polynomial mirror (M8) and the three high-density polyethylene (HDPE) corrugated lenses (L1, L2 and L3). The installation of the NIKA cryostat in the configuration presented before is shown in Fig. 9.1 including the mirrors M5, M6, M7 and M8.

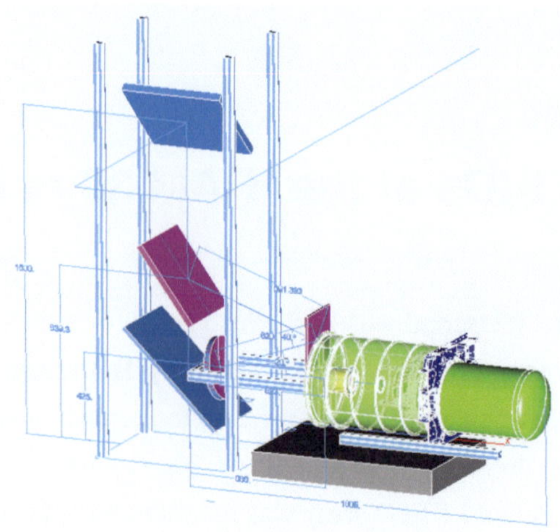

Fig. 9.1.: Installation schematic of the NIKA camera in the telescope cabin showing the cryostat and the final mirrors M5-M8, where the red arrow indicates the beam path.

9.1.1. The 2.05 mm LEKID array

A picture of the 144 pixel LEKID array used for the telescope engineering run is shown in Fig. 9.2. It is mounted in a copper sample holder for better thermalization reasons, where the back-short is integrated in the cover (not shown) of the holder with a distance of 700 μm. Due to the limited bandwidth of the readout electronics, only 98 pixels could be used for observations. These pixels are highlighted in white in the picture. Around 5 % of the resonators could not be measured due to fabrication errors, which are mainly those in the center area of the array. The focal plane geometry of each array is measured by using scanning maps of planets (see Fig. 9.3). The fitted focal plane geometry is found by matching the pixel position in the array as measured on the wafer to the measured position on planets, by optimizing a simple set of parameters: a center, a tilt angle and a scaling expressed in arcsec/mm. Here, most detectors are within less than 2 arcsec of their expected position. The beam width is also found from planet measurements. Typically the FWHM is around 16 arcsec for the 2.05 mm array with a dispersion of 1 arcsec.

Astronomical data from the two arrays are reduced offline with dedicated software. The raw data (I,Q) are converted to complex phase angle using the closest previous KID calibration. Then a conversion to an equivalent frequency shift is done with the same calibration using the derivative of the frequency with the complex phase at

Fig. 9.2.: 144 pixel LEKID array mounted in a copper sample holder. White colored pixels indicate the 98 usable resonators. Non colored represent the resonators that could not be used due to an exceeding of the readout electronic bandwidth of 230 MHz.

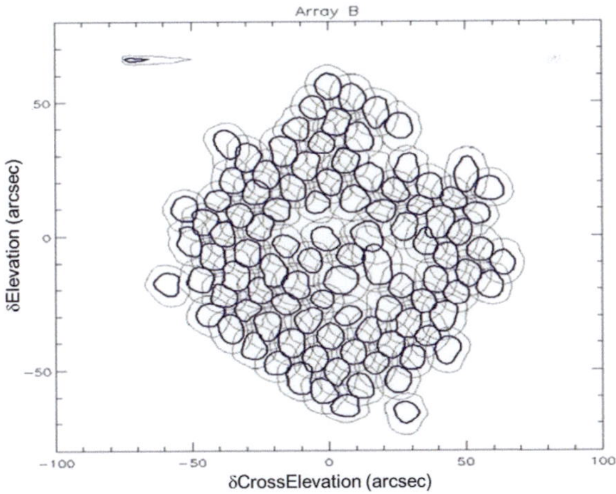

Fig. 9.3.: Contours of beams measured on Mars in October 2010 [44] of 98 valid pixels. Blue line: 50 % contour; grey line: 80 % contour.

the zero phase, as described in Chapter 8.2.4. Data are thus internally converted to frequencies which are assumed to be linear with the absorbed photon counts given by [59]

$$\Delta f_0 = -C f_0^3 \Delta L_K \propto -\frac{f_0^3}{n_s^2} \Delta P_{pixel} \tag{9.1}$$

where C is a constant.

9.1.2. Sensitivity measurement using planet Mars

The individual LEKID sensitivities are determined using a noise spectrum taken during an on-the-fly observation of the galactic source G34.5 following the same strategy as in the laboratory with the sky simulator. This noise spectrum is shown in Fig. 9.4, where the two contributions (low and high frequency fluctuations) can easily be separated. The low frequency noise is due to sky fluctuations and the expected source, whereas the fluctuations at higher frequencies are caused by the

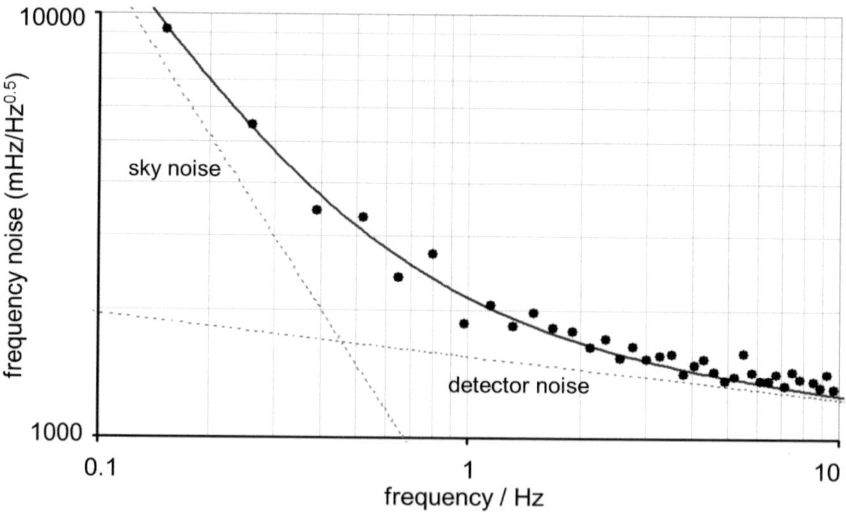

Fig. 9.4.: Frequency noise spectrum taken during a typical on-the-fly scan over the G34.3 galactic source [44]. The contributions of the sky and detector noise can easily be separated as indicated.

detector noise. A fit,using the sum of two power laws, to the data is given by

$$S_n(f) = A \cdot f^{\alpha} + B \cdot f^{\beta} \tag{9.2}$$

where $\alpha = -1.35$ at low frequencies and $\beta = -0.15$ at higher frequencies. This noise spectrum ,with an average value at 1 Hz of around 2 $Hz/\sqrt{Hz}$, corresponds to the results obtained in the laboratory. In addition, the relatively flat detector noise indicates an important reduction of the intrinsic frequency noise [23] compared to the first NIKA prototype tested at the telescope in 2009 [45].
Measuring the detector response during a scan on Mars, which is also used as a primary calibration source, and the noise spectrum from above, we can calculate the sensitivity of the LEKIDs under observation conditions. The power on one pixel has been determined to $P_{pixel} \approx 0.47\ pW$ taking the 50% transmission of the polarizer into account. With a response of around 4 kHz leading to a signal to noise ratio of $SNR = 2000\ \sqrt{Hz}$, we calculate an average optical noise equivalent power of $NEP_{optical} = 2.3 \cdot 10^{-16}\ W/\sqrt{Hz}$ per pixel, well in agreement with what we measured in the laboratory using the sky simulator.

9.2. Observed sources at 2.05mm

During the engineering run in October 2010, numerous objects have been observed, such as multiple and extended sources to demonstrate the mapping capability of NIKA. A series of faint sources has been observed to determine the sensitivity of the NIKA instrument on the sky. We show here only three of this large number of detected sources. The intensities of the here presented objects are given in Jansky (Jy), which describes the spectral flux density of an astronomical source with $1\ Jy = 10^{-26}\ W/Hz/m^2$. This non-SI unit is more appropriate to the intensity of the usually very weak sources and is therefore more practical to handle for such low power densities.
The first observed object we show here is a supernova remnant called *CassiopeiaA* or *CasA* as an example of a multiple source with extended diffuse emission (see Fig. 9.5A)). The contours of this source, observed at 2.05 mm are at levels of 32, 63, 95, 127 and 158 mJy/beam. The negative brightness features come from the filtering which is applied in order to reduce the sky noise.

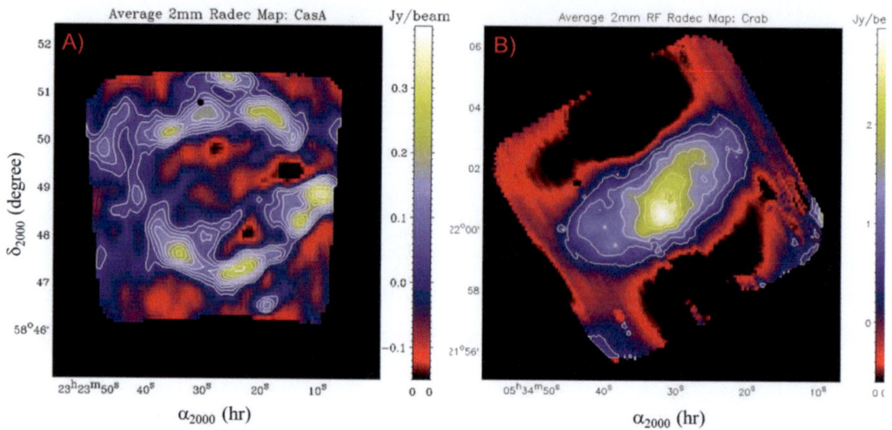

Fig. 9.5.: (A) Cassiopeia A (CasA), a supernova remnant and (B) the Crab nebula, a supernova remnant with pulsar wind nebula in the center.

In Fig. 9.5B we show the famous *Crab nebula* also known as *Messier* 1 (*M*1), a supernova remnant with a pulsar wind nebula in the center of the source, which can be found in the constellation of *Taurus*. The neutron star in the center emits pulses of radiation over a large spectrum, from gamma rays to radio waves, making this source interesting for testing new prototype instruments as NIKA. The Crab Nebula has recently been observed by GISMO (Goddard IRAM Superconducting 2 Millimeter Observer) [4], a 2 mm prototype instrument based on *Transition Edge Sensors* (*TES*). The obtained NIKA results are hereby well comparable with the GISMO observation, showing NIKA for the first time competitive with existing bolometer based instruments at the same wavelength. As an example of a relatively weak source and to demonstrate the camera capacity of detecting faint sources, we present in Fig. 9.6 a *Wolf-Rayet* (*WR*) radio star. The VCLS 147 has a measured flux of 162$\pm$12 mJy, which agrees with former observations [3]. The two symmetric negative lobes are an artifact of the filtering procedure mentioned above. *Wolf-Rayet* objects are super-massive stars ($\approx$20 times the solar mass) that are rapidly losing large amounts of masses by means of very strong stellar winds. Its characteristics are therefore very large emission lines that can be observed relatively easy. This source is one of the faintest that has been observed with NIKA in October 2010. The sensitivity, expressed in terms of flux density, is given by the *Noise equivalent flux density* (*NEFD*) and has been determined for NIKA at 2.05 mm

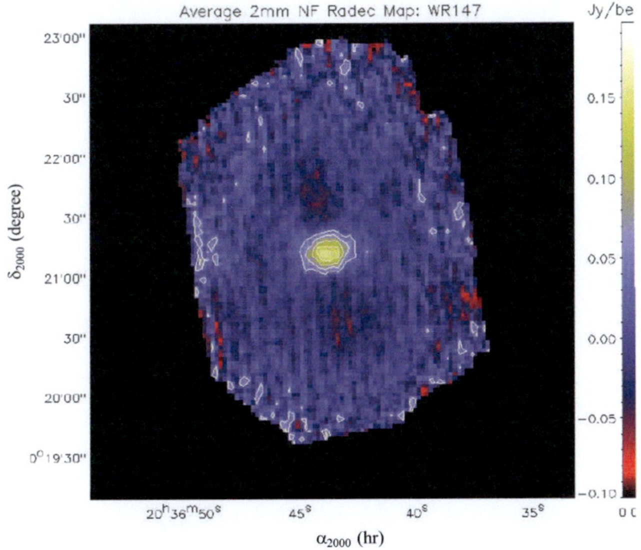

Fig. 9.6.: Map of *Wolf-Rayet* (*WR*) star VCLS147 with flux in the center of 162±12 mJy.

to $NEFD = 37\ mJy\sqrt{s}$. For comparison, MAMBO-2 reaches between 20 and 40 $mJy\sqrt{s}$ depending on the sky conditions [47]. The MAMBO-2 continuum instrument based on semiconductor bolometers is permanently installed at the IRAM 30 m telescope and observes at 1.2 mm with 128 pixels.

This second successful telescope run of NIKA with a dual band observing configuration has shown for the first time LEKID sensitivities of $NEP \approx 2.3 \cdot 10^{-16}\ W/\sqrt{Hz}$ at the 2.05 mm band and is therefore the today's most sensitive KID based instrument in this frequency range using the LEKIDs developed in the framework of this thesis. Other prototype instruments such as DEMOCAM [54] using antenna coupled MKIDs have shown their first light in 2008 but in the submillimeter range with sensitivities of around 1 $Jy/\sqrt{s}$.

10. Dual polarization LEKIDs

The LEKIDs discussed so far showed already excellent absorption properties and high sensitivity of around $NEP = 2.3 \cdot 10^{-16}\ W/\sqrt{Hz}$ per pixel using a polarizer in front of the array as described in the previous chapters. But the optical efficiency is still relatively low due to the single polarization properties of the meander line. To increase this efficiency for LEKIDs further, without changing the aluminum as superconductor, more optical power has to be absorbed in the LEKID structure. The sensitivity for the meander LEKIDs presented before was increased by optimizing the optical coupling and the kinetic inductance fraction. The kinetic inductance has reached its limits by reducing the volume of the meander lines to a minimum. It is difficult to deposit aluminum films thinner than 20 nm with an acceptable film homogeneity over the whole wafer. Also the aluminum gets transparent at mm-wavelength if the film is too thin leading to a decrease in optical coupling. Another limiting factor is the optical lithography. Long meander lines with w $< 3\ \mu m$ to decrease the line volume get difficult to achieve on large arrays. Alternative lithography methods using a laser aligner or an electron beam system takes too much time for the writing process and is not a solution for future large pixel arrays. One possibility to increase the optical efficiency is the use of a geometry that absorbs more power than the meander structure. The classic meander geometry is sensitive to only one polarization with little cross-polarization and cuts half of the power that could be absorbed. A LEKID geometry sensitive to two polarizations would thus increase the optical coupling leading to a higher responsivity of the detector. We present in this chapter one possible dual polarization LEKID design and discuss parameters such as resonator quality factors and optical coupling for this new geometry and verify them by simulations and measurements.

10.1. Hilbert curve as direct detection area

As demonstrated before, the absorptivity of the LEKIDs depends on the sheet resistance $R_{\square}$ and the filling factor defined by the ratio w/s. A constant filling factor is necessary over the whole direct detection area for a new LEKID design to guarantee a homogeneous optical coupling. To obtain this for a LEKID that is sensitive to two polarizations, a constant filling factor in both polarizations is needed. The structure should also be as symmetric as possible to absorb the same power in both polarizations with one common back-short. If the absorption in the two polarizations differs too much in frequency, two back-shorts would be needed for an optimal impedance matching, which is not realizable. In other words, the design must have comparable impedances for an incoming plane wave in both polarizations, which depends strongly on the symmetry of the inductive line.
Among many other geometries, $space-filling\ fractals$ such as the Peano or Hilbert curve [52] fulfill these conditions. These self-similar and scale invariant fractal patterns [35] are sometimes used as patch antennas for wireless communication systems [71, 64] and their antenna characteristics are well studied [70]. The antenna behavior is less important in our case because here we take advantage of the constant filling factor geometry to use it as a direct absorber with high mm-wave coupling efficiency in two polarizations.
Due to its symmetrical properties and the location of the start and end point of the curve, we decided to use the Hilbert curve as the most suitable fractal as an absorbing area for our purposes. Not only due to its symmetrical shape and constant filling factor, but also because it allows an easy integration of the interdigital capacitor the same way as for the classical meander structure, which would be more difficult using a Peano curve. The first six iterations of such a Hilbert space filling fractal is shown in Fig. 10.1. Higher order Hilbert curves are all based on the geometry of the first iteration (n = 1). The dimensions of the Hilbert curve are given by the length of each line segment h, the side dimension of the curve L and the iteration number or order n. The next higher iteration number (n = x + 1) can be considered as multiples of the iteration n = x by arranging them in different orientations. For example, the iteration n = 3 can be thought as four times the iteration n = 2. The

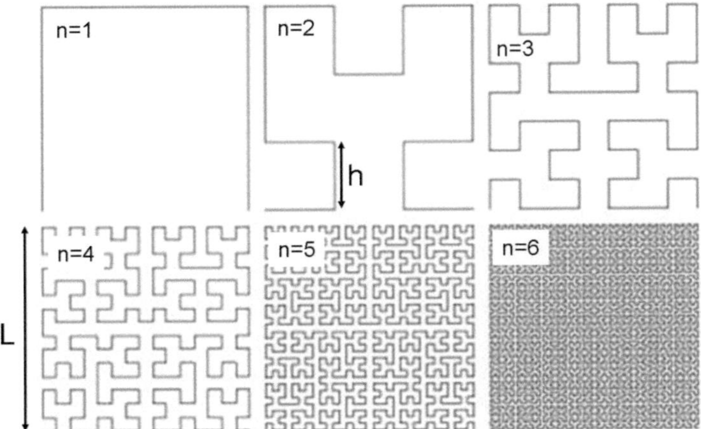

Fig. 10.1.: The space filling Hilbert fractal for the first six iterations ($n = 1-6$). The dimensions of the Hilbert curve are defined by the length of each line segment h and the side dimension L.

length of one line segment for a given dimension L can be calculated with [70] to

$$h = \frac{L}{2^n - 1} \tag{10.1}$$

where the total length of the curve line S is given by

$$S = (2^{2n} - 1)h \tag{10.2}$$

With these two simple equations and for a given pixel size, it is possible to calculate the entire Hilbert curve. The filling factor is defined by the iteration number n and the side dimension L and can therefore not be chosen randomly, which limits the impedance matching.

For a high optical coupling, the filling factor has to be sufficiently high to achieve a resistance R for optimal impedance matching (see Chapter 7). The filling factor of the optimized meander LEKID geometry $w/s = (92)^{-1}$ showed high optical absorption and is therefore taken for the new geometry. The length of the inductive line determines the resonance frequency of the resonator and should neither be too short, nor too long depending on the cross section. If the line is too short, the reso-

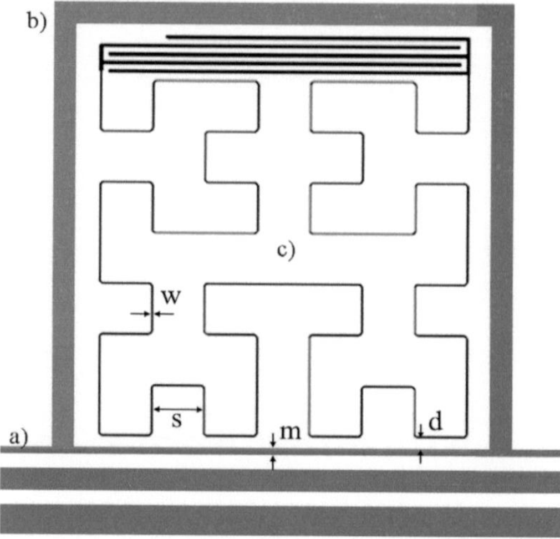

Fig. 10.2.: a) Coplanar wave guide as transmission line, where the resonator is inductively coupled to; b) surrounding ground plane frame to reduce pixel crosstalk; c) LEKID geometry using a Hilbert curve 3rd order as inductive part and an interdigital capacitor as capacitance of the resonant circuit (2mm^2).

nance frequency is too high due to a low inductance, which exceeds the bandwidth of the cryogenic amplifier. In addition, for a filling factor that is too low, the resistance R of Z_{LEKID} is too high leading to a mismatch in optical coupling and thus to less absorption. This excludes Hilbert curves with $n < 3$ and $n > 3$ in the case of aluminum resonators. For materials with higher resistivity, a higher filling factor is possible to achieve the same resistance R.

A solution for an aluminum LEKID geometry with acceptable filling factor and sufficient line length S is the Hilbert curve 3rd order. Keeping the same pixel size as for the meander LEKID this allows a geometry with comparable filling factor in both polarizations. In Fig. 10.2 the design of a Hilbert LEKID coupled to a cpw transmission line is shown. We keep the ground plane frame from the original design for the same reason, to reduce the cross coupling between pixels. The design has a filling factor of $\mathrm{w}/s = (92)^{-1}$ with w$= 3\ \mu m$ and $s = 276\ \mu m$.

Of course the filling factor is not exactly as the one from the meander geometry due to the interrupted lines in horizontal and vertical orientation. The spacing between these interruptions is smaller than the wavelength, and we assume here, that for an incoming plane wave with $\lambda_0 = 2.05\ mm$, the interrupted lines are considered as connected. This assumption has of course to be verified by simulations and measurements and will be discussed later. The pixel is slightly bigger in height due to the squared shape of the Hilbert curve, where the capacitor is added to. The interdigital capacitor dimensions are unchanged in line width (10 μm) and distance between the lines (20 μm). The inductive line is around 5 % longer than that of the meander LEKID, but this only leads to a slightly lower resonance frequency. The properties of this new LEKID design have been characterized electrically (coupling to transmission line, quality factors and resonance frequency) and optically (absorption measurements and numerical simulations). The coupling to the transmission line has to be adapted for the new geometry to optimize the quality factors of the resonator. Also the frequency tuning has to be simulated to achieve equally spaced resonances in a limited bandwidth as it is done for the classical meander geometry. The most important part of this characterization is the measurement of the optical coupling and the comparison to the classical meander LEKID.

10.2. Resonator properties at 4 K

As for the meander design, the coupling of the Hilbert LEKID to the transmission line has to be optimized regarding a high sensitivity of the detector. This has been simulated using *Sonnet* and verified by low temperature measurements. We designed therefore a 25-pixel array, which is shown in Fig. 10.3 containing five different couplings defined by m and d (see Fig. 10.2). We have chosen a resonance frequency spacing of 5 MHz between the pixels and 10 MHz between the lines with different coupling in order to minimize cross coupling effects. For the presented geometry this corresponds to 1 MHz/12 μm of the capacitor line. To determine the frequency tuning, a single resonator has been used for the simulation in order to reduce simulation time. This does not take into account any cross coupling effect, but the simulation with sufficient frequency resolution of a 25 pixel array would be too time consuming. The measurement has been performed using Nb as superconductor due to its higher T_C, which is less time consuming and the electrical properties are well comparable to aluminum. The internal quality factor Q_0 is of course different because it depends directly on the temperature and the surface resistance. Here we concentrate on the optimization of the coupling to the transmission line and Nb can therefore be used to determine Q_C. The measurement of the 25-pixel

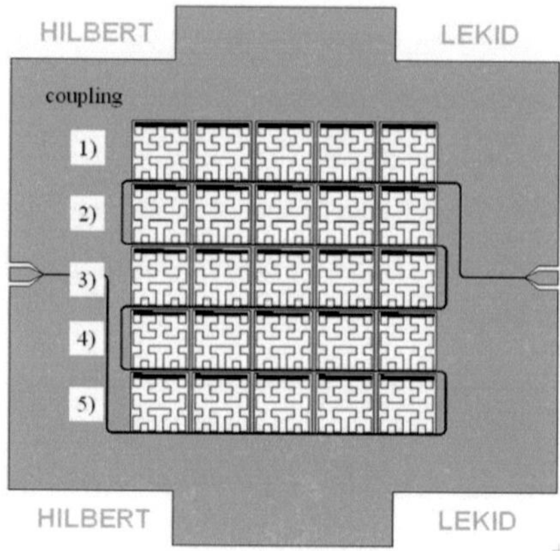

Fig. 10.3.: Design of a 25-pixel Hilbert-LEKID array with 5 different couplings of the resonator to the transmission line. 1) $m\ [\mu m] = 10/d\ [\mu m] = 14$, 2) $12/14$, 3) $12/12$, 4) $14/12$, 5) $16/16$.

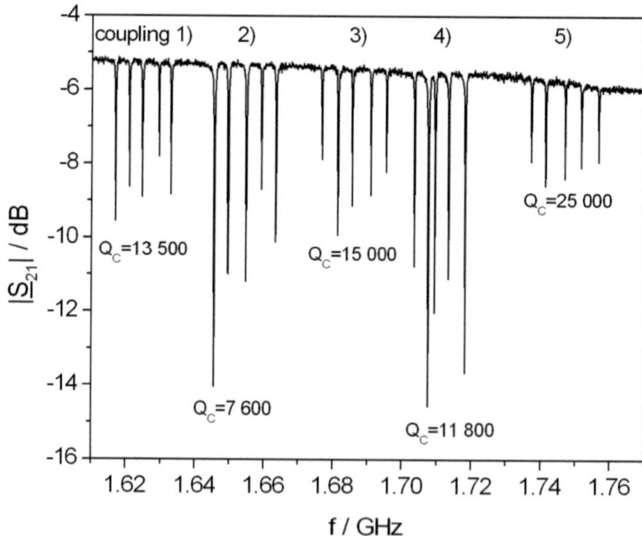

Fig. 10.4.: Measurement of a 25-pixel Hilbert LEKID array for five different couplings to the transmission line. The array is made of a 60 nm niobium film dc-sputtered on a silicon substrate and measured at 4.2 K in liquid helium.

array made of Nb on a silicon substrate is shown in Fig. 10.4. This first measurement of Hilbert-LEKIDs confirms the feasibility as resonant circuit for KIDs with comparable resonator properties. The resonances are equally spaced in frequency and agree with the simulated spacing. The different shape of the line seems to have no influence on the electrical properties of the resonator. The resonance frequencies are at around 1.7 GHz, which would be lower for aluminum due to the higher kinetic inductance. The calculated coupling quality factor Q_C varies from 7600 to 25000 and is therefore comparable to the meander LEKID, which is due to the same geometry at the coupling area. From measurements and simulations of the meander LEKID we know, that a Q_C between 25000 and 35000 is sufficient to guarantee high sensitivity and dynamic range. Thus, for the further study of this new structure, the coupling 5) of Fig. 10.4 is used corresponding to the same coupling of the meander LEKID with $d = 16\ \mu m$ and $m = 16\ \mu m$.

10.3. mm-wave coupling of Hilbert LEKIDs

The new dual polarization LEKID is optically coupled the same way as the meander LEKID, illuminated from the backside of the substrate and a back-short cavity mounted at a calculated distance in front of the sample. A schematic of this including substrate and back-short cavity is demonstrated in Fig. 10.5. The absorptivity

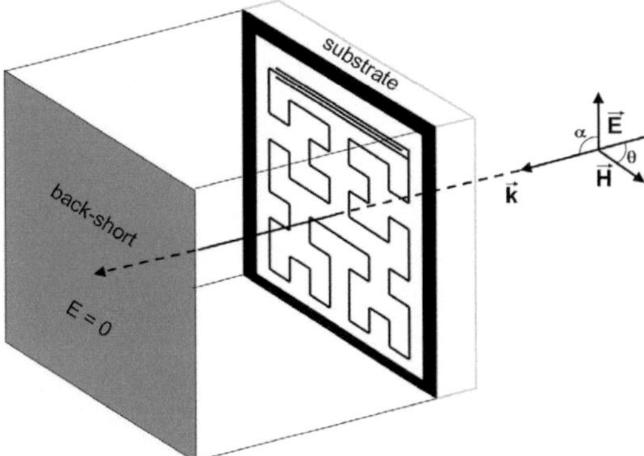

Fig. 10.5.: Schematic of the optical coupling. The LEKID is illuminated from the backside of the substrate. At a calculated distance to the substrate, a back-short cavity is situated to increase the mm-wave absorption.

here is also optimized by adapting the back-short to the impedance of the Hilbert LEKID, where the substrate thickness is kept unchanged at $l_{sub} = 300\ \mu m$. The optimization of the optical coupling has been done using numerical simulations and room temperature reflection measurements as described in Chapter 7. We present in this section the results of these measurements and compare the absorptivity to that of the classical meander geometry.

10.3.1. Measurement of the mm-wave absorption at room temperature

In order to measure the absorption in both polarizations, we mounted the sample on a rotatable holder. The measurements have been performed with the capacitor lines perpendicular and parallel to the electrical field orientation of the incoming plane wave. In Fig. 10.6 we show the optical absorption of the Hilbert-LEKID, calculated from the measured reflection scattering parameter S_{11} using a back-short length of

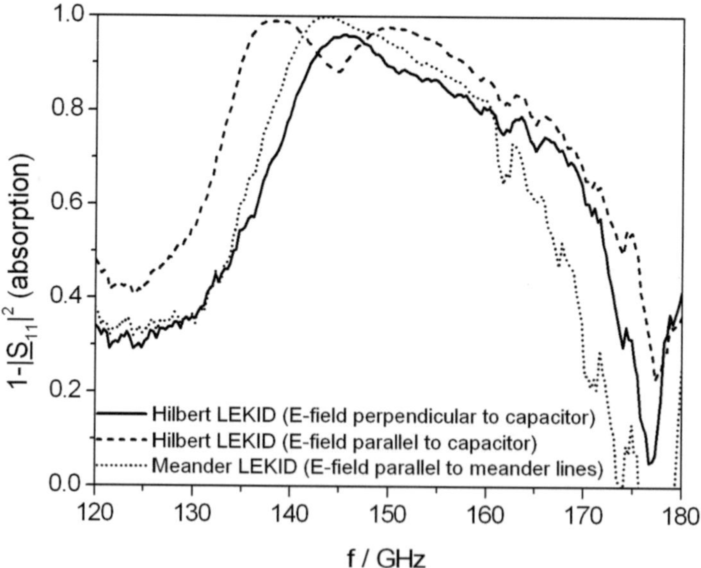

Fig. 10.6.: Measurement of the optical absorption of the Hilbert LEKID geometry for two polarizations (electrical field parallel or perpendicular to the capacitor lines). The back-short of the Hilbert-LEKID is 600 μm and that of the meander LEKID structure 800 μm.

$l_{bs} = 600\ \mu m$. This length was determined by measuring the sample with different back-shorts and comparing the absorptivity over the frequency bandwidth (120-180 GHz). The absorption is shown for both polarizations and compared to the classical meander LEKID with same filling factor and surface resistance. Regarding the absorption in one polarization, it is comparable to that of the meander geometry. This confirms the assumption, that the interrupted lines can be considered as connected if the distance between them is smaller than the wavelength. This becomes more obvious by comparing the measurements of the Hilbert- and the meander LEKID with the same pixel orientation, where the capacitor lines are perpendicular to the incoming plane wave (comparing dotted and solid line in Fig. 10.6). Here we see only a small difference in absorption, which means, that the long parallel meander lines and the interrupted parallel lines have comparable filling factors. The back-short of the Hilbert-LEKID is shorter for an optimal coupling, which means, that the inductance L of Z_{LEKID} is smaller, but the absolute absorption of almost 100 % indicates a comparable resistance R. If R is comparable in both cases, the filling factor also is. The results from this measurement also confirm the expected improvement in optical coupling due to the dual polarization geometry of the Hilbert-LEKID. It shows an absorption, taking both polarizations into account that is around two times higher compared to the meander design. From this we can therefore expect a responsivity that is also around two times higher. The absorption in the second polarization, where the capacitor lines are parallel to the electrical field, is slightly different. This is due to the long capacitor lines that absorb some power in this orientation. The filling factor is here too high and the absorption in this part of the geometry is relatively low compared to the rest of the LEKID but not zero.

10.3.2. Numerical simulations of the absorptivity

To determine the optical coupling by numerical simulations, we use the same CST model as for the meander LEKID, where a plane wave is generated by defining the boundary conditions of the waveguide model as described in Chapter (7.3). To simulate the absorption in both polarizations, two independent simulations have been performed, with two different orientations of the LEKID structure in the waveguide. One with a perpendicular and one with a parallel orientation of the capacitor lines to the electrical field, as it was done for the reflection measurements. The results of these simulations, for a back-short of 600 μm and a substrate thickness of 300 μm, are shown in Fig. 10.7 and 10.8 and compared to the absorption calculated from the reflection measurement. The simulated model includes the ground plane frame around the pixel with $a = 200\ \mu m$. As seen from the meander LEKID, a ground plane of this width has little influence on the total absorption letting us assume that most of the power is absorbed in the inductive line rather than in the ground plane. We see an excellent agreement between simulation and measurement in both polarizations, allowing us to use this model to optimize the optical coupling for these or similar LEKID geometries.

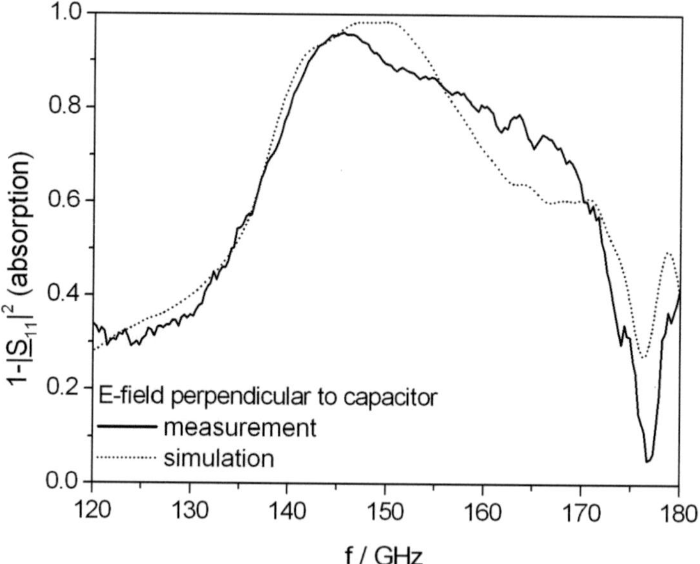

Fig. 10.7.: Comparison of simulation (dashed line) and measurement (solid line) of the Hilbert-LEKID geometry for a plane wave with electrical field perpendicular to the capacitor lines. The back-short distance is 600 μm.

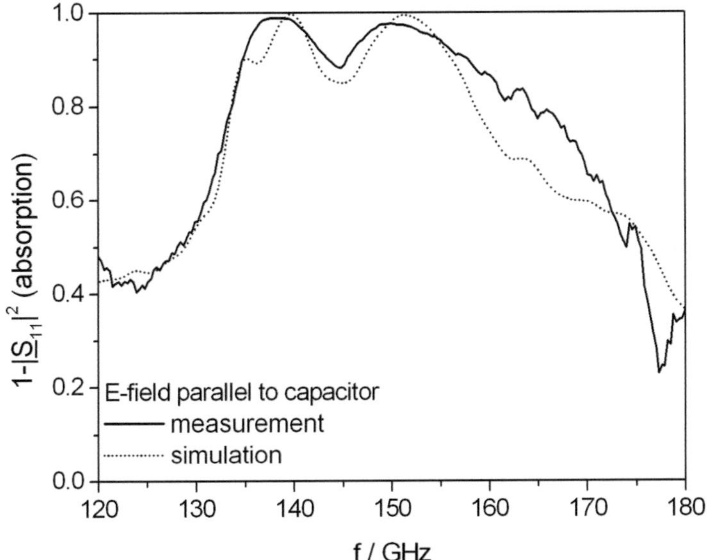

Fig. 10.8.: Comparison of simulation (dashed line) and measurement (solid line) of the Hilbert-LEKID geometry for a plane wave with electrical field parallel to the capacitor lines. The back-short distance is 600 μm.

10.4. Optical characterization in the NIKA camera

Based on these results, a 132-pixel Hilbert-LEKID array has been designed and fabricated to be measured in the NIKA cryostat. The results of these measurements are discussed in the following section.

10.4.1. Cryostat upgrade

To measure samples in the presented NIKA camera that are sensitive to two polarizations, we needed to change one part of the optics in the cryostat. We could not use a polarizer anymore to split the signal for the two frequency bands to measure the absorption in both polarizations. For that reason, the polarizer is replaced with a dichroic filter that allows splitting the frequency band without losing one of the polarizations. This filter is mounted as the polarizer with an incident angle of 45 degrees. Depending on its orientation, the filter acts as low pass for the reflected

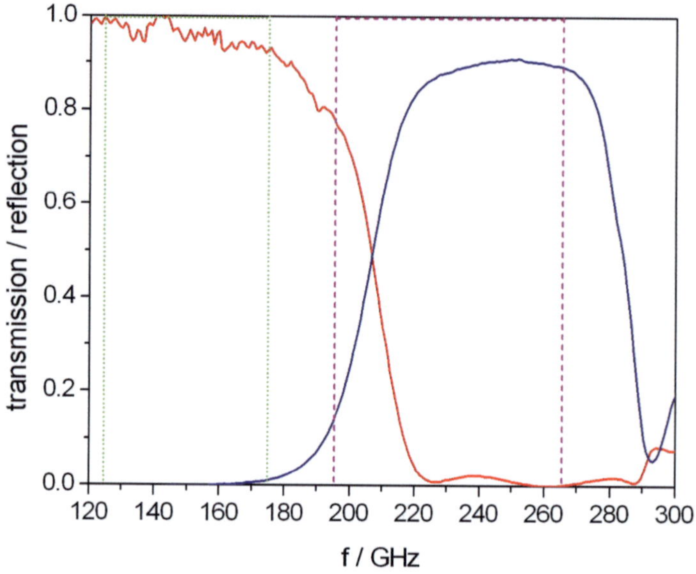

Fig. 10.9.: Transmission (blue line) and reflection (red line) spectrum of the dichroic filter in the NIKA cryostat compared to the spectra of the NIKA filters for the two bands (green dotted line and pink dashed line).

signal and as high pass for transmitted signals. In our case, the filter reflects signals for $f < 207Ghz$ to the 2.05 mm array and transmits signals with $f > 207GHz$ to the 1.25 mm array. In Fig. 10.9 we present the frequency dependent transmission and reflection properties of the used dichroic filter. This spectrum was measured for an incident angle of 45 degrees and we can see a certain overlap between the two bands. The used filter was originally designed for an angle of 35 degrees, and we lose therefore in filter efficiency in particular at the 1.3mm band. The spectrum is compared to the -3 dB bandwidth of the filters inside the camera, presented in Chapter 8 that provide the frequency band of the two arrays. The spectrum shows that we are not limited by the dichroic filter at the 2.05 mm band and can count the whole bandwidth for the characterization of the Hilbert LEKIDs. At 1.25 mm on the other hand, we lose around 20GHz of bandwidth due to the dichroic filter, which has to be taken into account for the characterization of the pixel sensitivity of the 1.25 mm KID array.

10.4.2. Characterization of Hilbert LEKIDs at T=100 mK

Responsivity in different polarizations To verify the optical response of the Hilbert LEKIDs, the sky simulator was cooled down to a temperature of 210 K allowing to measure with a $\Delta T = 300\,K - 210\,K = 90\,K$. This relatively high temperature difference is chosen to achieve a high detector response in order to distinguish between the responsivities in the different polarizations. To determine the detector response independently in both polarizations, we mounted a polarizer in front of the cryostat window to compare the single polarization response to the total response (without polarizer). Using a network analyzer, we measure the S_{21} transmission parameter of one resonator and compare the shift in resonance frequency for the different optical loads. The response is measured for two different polarizer orientations, horizontal (PH) and vertical (PV) to the orientation of the interdigital capacitor lines of the KID. The results of these measurements are presented in Fig. 10.10 for one resonator of the 132 pixel array. We measure an identical KID response of $\Delta f \approx 38\,kHz$ for a temperature difference of $\Delta T = 90\,K$ using the polarizer in horizontal or vertical orientation. This confirms the results obtained from the room temperature reflections measurements and simulations, where they have shown comparable absorption properties of the Hilbert LEKID geometry in both polarizations in the 2.05 mm band. For the same measurement but without the polarizer in front of the cryostat we measure a KID response of around $\Delta f \approx 75\,kHz$ for the same ΔT corresponding to twice the single polarization response. If we compare the responsivity of the new Hilbert LEKID geometry (≈830 Hz/K) to the classical meander structure (≈445 Hz/K), we gained a factor of almost two. These results show, that the symmetrical geometry of the Hilbert curve with constant filling factor is suitable as a dual polarization LEKID and confirm the expected absorption properties due to a comparable impedance in both polarizations.

Sensitivity of the Hilbert LEKIDs An optical load of 210 K is relatively high and can decrease the intrinsic quality factor and thus the sensitivity. For better comparison reasons, we cooled the sky simulator down to 50K in order to measure under the same conditions as it was done for the meander LEKID. The sensitivity is determined the same way, by measuring the noise spectrum of each pixel to calculate the optical NET using equation (8.14). The measurement showed a si-

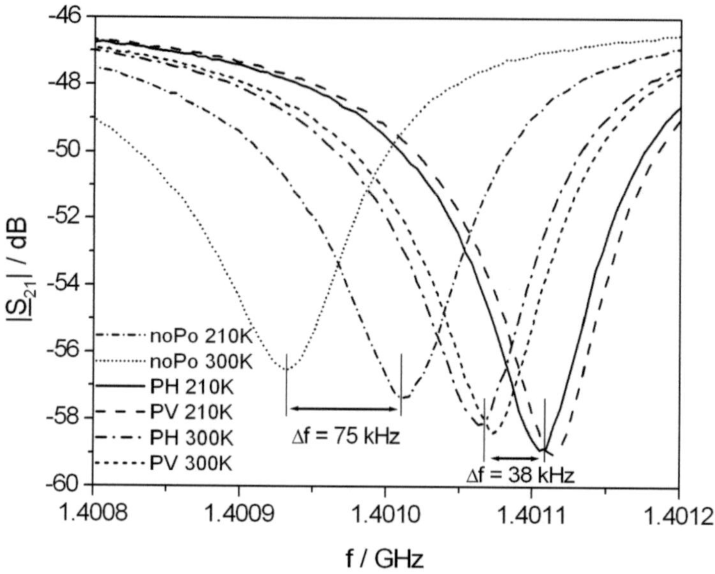

Fig. 10.10.: Measurement of the resonance frequency of a Hilbert LEKID for two different optical loads (210 and 300 K). It has been measured for the two polarizations individually using a polarizer in front of the cryostat and for both together without polarizer. PH: polarizer horizontal; PV: polarizer vertical; noPo: without polarizer.

milar noise spectrum as for the classical LEKID with $S_n(1Hz) = 2\ Hz\sqrt{Hz}$. The higher power on one pixel due to the dichroic filter has not increased the noise, which means that the LEKIDs have not reached the photon limit yet. Increasing the temperature by $\Delta T = 5\ K$ gives a signal to noise ratio of $SNR = 2400 \cdot \sqrt{Hz}$ with a responsivity of 960 Hz/K leading to an average noise equivalent temperature of $NET_{optical} \approx 2.1\ mK/\sqrt{Hz}$. From this result we see, that the responsivity under comparable background conditions has been increased by more than a factor of two and we calculate therefore a NET that is more than two times lower compared to the classical LEKID. Fig. 10.11 shows the NET calculated for 80 LEKIDs of the 132 pixel array, where the blue dashed line indicates the average optical NET.

If we now calculate the corresponding noise equivalent power, we find $NEP_{optical} = 2.1 \cdot 10^{-16}\ W/\sqrt{Hz}$. This value has only slightly improved compared to the meander LEKID. The reason for this is the different power that illuminates the pixel in

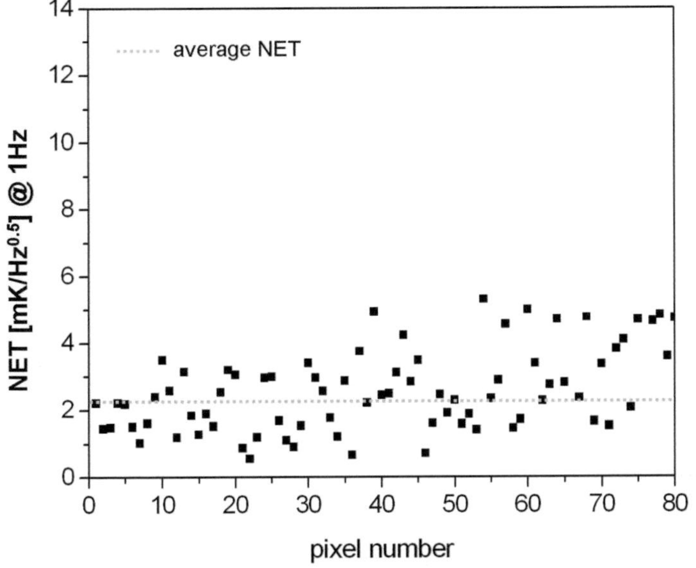

Fig. 10.11.: Noise equivalent temperature of 80 Hilbert LEKIDs measured at $T = 100\ mK$. The average NET is 2 mK/Hz$^{0.5}$ at 1 Hz per pixel (dashed line).

both cases. In the classical LEKID case we divided the pixel power by a factor of two due to the used polarizer. Here in the Hilbert case, where a dichroic filter is used to split the signals, twice the power is illuminating one pixel. A factor of two in responsivity improves the NEP of course, but a factor of two in optical power results in the same value (see equation 8.12). If the dichroic filter had been used for the classical LEKID, the NEP would be higher by a factor of two for the same responsivity due to the absorption properties of the meander line in only one polarization. These Hilbert LEKIDs have successfully been tested during a third NIKA telescope run in October 2011. The data analysis of these results is ongoing and more detailed information can therefore not be given here. Preliminary data confirm the results from the laboratory and indicates an improvement in sensitivity compared to the test run in 2010 by a factor of 1.5 in NEP. The Hilbert LEKIDs have therefore proved their applicability as highly sensitive kinetic inductance detector for radioastronomy (in particular at 150 GHz) with improved sensitivity and optical efficiency compared to the classical meander geometry. Since these obtained results, the Hilbert LEKID is used as detector at 2.05 mm for the NIKA instrument and replaced the classical meander LEKID.

11. Conclusion and perspectives

This thesis presented the development, the fabrication and characterization of so-called *Lumped Elements Kinetic Inductance Detector* (*LEKID*) arrays as mm-wave continuum detectors. This work arose within the framework of the NIKA (*Néel – IRAM KID Array*) collaboration, which is the current development of a large field-of-view camera for the IRAM 30 m telescope in Spain observing simultaneously at 1.25 and 2.05 mm (240 and 146 GHz) based on *Kinetic Inductance Detectors* (*KID*).
To prove the large scale multi-pixel suitability and high direct mm-wave absorption properties of such detectors, the following points have been subject of detailed studies throughout this thesis.

- Design and modeling of single superconducting aluminum microresonators and optimization of the according LEKID parameters to meet the NIKA specifications.
- Development of multi-pixel LEKID arrays and investigation of parameters such as resonance frequency tuning, resonator packaging density and electromagnetic cross coupling effects.
- Thin film deposition and microprocessing to build LEKID arrays.
- Modeling and optimization of the mm-wave coupling efficiency of LEKIDs in the 2.05 mm band (125-175 GHz), including the proposal and verification of a new method to determine the mm-wave absorption using a quasi-optical reflection measurement setup at room temperature.
- Characterization of LEKIDs at sub-Kelvin temperatures in order to determine the detector's sensitivity.
- Design of a dual polarization LEKID geometry in order to achieve higher mm-wave absorptivity.

The particular suitability of LEKIDs for use as large filled arrays has been demonstrated by designing and measuring multi-pixel arrays and optimizing the according resonator parameters, such as coupling and intrinsic quality factors, the resonance frequency tuning and a homogeneous frequency spacing. The resonators have been designed as a compromise between coupling and loaded quality factor in order to achieve detectors with high sensitivity and sufficient dynamic range. One of the most disturbing parameters, the electromagnetic pixel cross coupling, has been reduced by significant values due to a ground plane shielding between the detectors and a particular geometrical resonator distribution, allowing the design of even larger arrays.

One big advantage of this type of detector is the relative easy fabrication process. Even for large arrays, a single superconducting thin layer, sputtered or evaporated, on a silicon substrate suffice to fabricate the LEKIDs within less time and less expensive compared to the relative complex readout circuits of TES. The importance of a homogeneous thin film deposition process has been demonstrated, which has high impact on the design of large detector arrays due to a direct dependence of the kinetic inductance on the film thickness. Variations in film thickness over the surface of a 2 inch or even bigger substrate makes it difficult to design resonators with equally spaced resonances in a limited bandwidth.

Another important factor for this direct absorption detector type is the modeling and optimization of the mm-wave coupling properties. A quasi-optical room temperature reflection measurement setup has been proposed, allowing to measure the mm-wave absorptivity of LEKIDs without cooling them down to sub-Kelvin temperatures. The results of these measurements have been validated by analytical calculations and numerical simulations, showing excellent agreement. The presented setup gives the possibility to characterize future LEKID geometries at room temperature in less time with comparable results to the cryogenic case. The tested LEKIDs showed high absorptivities in the 2.05 mm band, corresponding to frequencies between 125 and 175 GHz, with maximum absolute absorption up to almost 100 %.

The developed LEKIDs have been optically characterized at low temperatures, using a cryostat with optical access and base temperature of $T = 100\ mK$ in order to determine the sensitivity. They have shown sensitivities in the range of $NEP_{optical} = 2 \cdot 10^{-16}\ W/Hz^{0.5}$, which corresponds to a *N̲oise E̲quivalent F̲lux*

Density (*NEFD*) of $NEFD = 37\ mJys^{0.5}$. This is the first time that LEKIDs, in the 2.05 mm band, have shown such high sensitivity. These numbers have been confirmed during a successful engineering run at the IRAM 30 m telescope, where the NIKA LEKIDs not only showed the feasibility of mapping extended galactic objects, but also the observation of faint sources. With the detector development done during this thesis, the imaging qualities of LEKID arrays have been demonstrated for the first time. The detector properties have been improved even further, making them today's most sensitive KID based detectors at 2.05 mm (146 GHz). The designed and optimized LEKIDs are now competitive with existing bolometer instruments such as MAMBO (semiconductor based detectors) and GISMO (TES), proving their ability for high sensitivity detectors for mm-wave astronomy.

A novel dual polarization geometry for the NIKA LEKIDs based on the face filling Hilbert curve has been developed and tested. This LEKID geometry showed high and comparable absorption properties in both polarizations leading to an improvement in optical efficiency by a factor of two compared to the classic meander LEKID. It has recently successfully been tested during a third telescope engineering run, where preliminary data indicate an significant improvement in detector sensitivity.

The presented work is basis of a future project: an upgrade of the NIKA instrument. The pixel count for the 2.05 mm array will be around 1000 pixels, which is a big challenge, in particular concerning the fabrication process and pixel cross coupling effects. For this array size, larger substrates of at least 4 inch in diameter are needed making a homogeneous film deposition and microfabrication processes more difficult. The first prototype of such an array is illustrated in Fig. 11.1 consisting of 1020 Hilbert LEKIDs coupled to four separate transmission lines. The design of this array is based on the results achieved during this thesis, including an improvement in pixel cross coupling and optical efficiency.

Based on the achieved results of the NIKA LEKIDS, the collaboration decided to develop also arrays for the 1.25 mm band (210-290 GHz). First arrays have been designed and tested using the Hilbert LEKID geometry, which have shown comparable results making them subject of ongoing research.

Fig. 11.1.: The picture shows a first prototype 1000 pixel array for the NIKA 2.05 mm band. The detectors used in the array are Hilbert LEKIDs, developed during this thesis.

12. Zusammenfassung und Ausblick

In dieser Arbeit wurden die Entwicklung, Herstellung und Charakterisierung von sogenannten *Lumped Elements Kinetic Inductance Detektoren* (*LEKID*) zur Detektion von Millimeterwellen vorgestellt. Sie entstand im Rahmen einer Kollaboration mit dem Namen NIKA (*Néel – IRAM KID Array*), deren Ziel es ist, eine Kamera mit großem Sichtfeld für das IRAM 30 m Teleskop zu entwickeln. Diese Kamera soll dafür ausgelegt sein, basierend auf *Kinetic Inductance Detektoren* (*KID*), Beobachtungen sowohl bei 146 als auch 240 GHz (2.05 and 1.25 mm) simultan durchzuführen.
Zum Prüfen der Eignung der LEKIDs als mögliche Detektoren für die Radioastronomie, insbesondere mit Blick auf Multi-Pixel-Anwendungen und Absorptionseigenschaften von Millimeterwellen, wurden in dieser Ausarbeitung die nachstehend genannten Untersuchungen verschiedener Parameter realisiert:

- Design und Modellierung von supraleitenden Mikroresonatoren aus Aluminium und Optimierung der entsprechenden Parameter zum Erfüllen der NIKA-Spezifikationen.
- Entwicklung von Multi-Pixel-Arrays mit Fokus auf den Resonanzfrequenzen, der Packungsdichte und den Effekten aufgrund von elektromagnetischem Übersprechen zwischen den Resonatoren.
- Herstellung und Mikroprozessierung von LEKID-Arrays.
- Modellierung und Optimierung der Einkopplung von Millimeterwellen in LEKIDs bei 125 bis 175 GHz (2.05 mm Band). Diesbezüglich wurde eine Methode zur Messung der Absorption mittels eines quasi-optischen Reflektionsmessaufbaus bei Raumtemperatur vorgestellt.
- Charakterisierung der LEKIDs bei kryogenen Temperaturen zur Bestimmung der Detektorempfindlichkeit im Labor wie auch am Teleskop.

- Design einer zweipolaren LEKID-Geometrie zur Erhöhung der Millimeterwellenabsorption im Detektor.

Die Eignung von LEKIDs als Detektoren für radioastronomische Beobachtungen wurde in dieser Arbeit erfolgreich demonstriert. Anhand von Simulationen und Messungen konnten die entsprechenden Parameter der Resonatoren, die für Multi-Pixel-Anwendungen relevant sind, bestimmt und optimiert werden. Dazu gehören die Güten der Resonatoren, ein äquidistantes Durchstimmen der Resonanzfrequenzen sowie die Minimierung des elektromagnetischen Übersprechens. Die Güten der Resonatoren wurden dementsprechend entworfen, um einen akzeptablen Kompromiss zwischen Empfindlichkeit und Detektordynamik zu erreichen. Einen der störenden Faktoren bei großen Arrays stellt das elektromagnetische Übersprechen dar, das mittels einer abschirmenden Massefläche zwischen den Pixeln signifikant reduziert werden konnte. Die hieraus generierten Erkenntnisse erlauben es zukünftig Arrays mit noch höherer Pixelanzahl zu konzipieren. Es konnte ferner gezeigt werden, dass die relativ simple Herstellungsweise der LEKIDs einen großen Vorteil gegenüber anderen Detektorkonzepten, wie z.B. TES, impliziert, die in der Regel sehr komplexe Mikroprozessierungsschritte benötigen. Dadurch lassen sich LEKIDs nicht nur mit geringem Zeitaufwand herstellen, sondern sie sind zudem sehr kostengünstig. Die Wichtigkeit von homogenen Schichtdicken des Aluminiums hat sehr großen Einfluss auf die kinetische Induktivität, weshalb ein kontrollierter Schichtabscheidungsprozess die Basis eines hochempfindlichen Detektors bildet. Dies wurde nicht nur rechnerisch, sondern gleichsam mittels kryogener Messungen demonstriert. Ein weiterer essentieller Parameter sind die Modellierung und Optimierung der Einkopplung von Millimeterwellen in den Detektor. Hierzu wurden ein analytisches Transmission Line Modell, ein 3D-Simulationsmodell und ein quasioptischer Messaufbau vorgestellt. Der gezeigte Messaufbau erlaubt die Bestimmung der Absorption von Millimeterwellen, ohne den Detektor auf kryogene Temperaturen herunter zu kühlen, was üblicherweise sehr viel Zeit in Anspruch nimmt. Simulationen und Messungen zeigten hier eine exzellente Übereinstimmung mit einer erzielten absoluten Absorption von nahezu 100%. Für die kryogene Charakterisierung der LEKIDs bei T = 100 mK wurde ein Mischkryostat mit optischen Zugang verwendet. Es wurden Empfindlichkeiten, gegeben in Noise equivalent Power (NEP), von $NEP_{optisch} = 2 \cdot 10^{-16}\ W/Hz^{0.5}$ erreicht, was einer Noise equivalent Flux density (NEFD) von $NEFD = 37\ mJys^{0.5}$ entspricht. In diesem

Frequenzbereich wurden bisher keine höheren Empfindlichkeiten von LEKIDs gemessen. Diese Werte wurden während eines erfolgreichen Tests am IRAM 30 m Teleskop bestätigt. Mit den in dieser Arbeit entwickelten Detektoren wurde erstmals die radioastronomische Bildgebung mit LEKIDs demonstriert. Die LEKIDs bewiesen hierbei ihre Fähigkeit, sowohl ausgedehnte galaktische Objekte zu mappen, als auch sehr schwach abstrahlende Quellen zu detektieren, wodurch NIKA das aktuell empfindlichste auf KID basierende Instrument in der Radioastronomie bei 2.05 mm ist. Die mit NIKA erzielten Resultate sind mittlerweile vergleichbar mit existierenden Instrumenten wie MAMBO (halbleiterbasierte Bolometer) oder GISMO (TES-basierte Bolometer), womit LEKIDs die Spezifikationen als hochempfindliche Detektoren für die Radioastronomie erfüllen. Darüber hinaus wurde eine neuartige zweipolare LEKID-Geometrie zur weiteren Steigerung der Millimeterwellenabsorption im Rahmen dieser Arbeit entworfen und charakterisiert. Diese Geometrie, die auf der flächenfüllenden Hilbert-Kurve basiert, zeigte eine annähernd identische Absorption in horizontaler wie auch vertikaler Polarisation, was zu einer Verbesserung der optischen Effizienz um einen Faktor 2 entspricht (verglichen mit der klassisch mäandrierten Geometrie). Ein Array mit dieser LEKID-Geometrie wurde ebenfalls während eines Tests am Teleskop eingesetzt und erste Daten deuten auf eine signifikante Verbesserung der Detektorempfindlichkeit hin.

Die vorgestellte Arbeit fungiert als Basis eines zukünftigen Projektes: In diesem wird es um die Erweiterung der NIKA-Kamera gehen. Hierbei soll die Anzahl der Pixel bei 2.05 mm auf ca. 1000 erhöht werden, was insbesondere in Bezug auf die Herstellung der Detektoren und das elektromagnetische Übersprechen eine Herausforderung sein wird. Für eine solche Pixelanzahl müssen größere Substrate (> 4") verwendet werden, was eine homogene Schichtabscheidung und Mikroprezessierung erschwert. Ein erster Prototyp eines solchen Arrays ist in Bild 11.1 zu sehen, bestehend aus 1020 Hilbert LEKIDs, die an vier separate Durchgangsleitungen angekoppelt sind. Der Entwurf des Arrays basiert auf den Ergebnissen, die in dieser Arbeit erzielt wurden, einschließlich einer Verbesserung des Übersprechens und der optischen Effizienz. Aufgrund der positiven Ergebnisse von NIKA wird nun auch die Entwicklung von LEKIDs für den Frequenzbereich von 210 bis 290 GHz (1.25 mm) angestrebt. Erste Arrays sind bereits entworfen und gemessen worden. Angesichts der ersten Ergebnisse, die sehr vielversprechend ausgefallen sind, werden diese das Subjekt weiterer Forschung sein.

A. Appendix

A.1. Fundamentals of superconductivity

The London theory The brothers F. and H London have been the first who described the basic electrodynamical properties of superconductors in 1935 using a phenomenological model for the conductivity below the critical temperature T_C [33]. To derive the London model for superconductors, we have to take a look to the standard Drude model for electrical conductivity for normal conductors. The classical mechanics of an electron motion due to an applied electrical field is described in this model by the equation of motion

$$m\frac{\partial \vec{v}}{\partial t} = e\vec{E} - \frac{m\vec{v}}{\tau} \tag{A.1}$$

where m and e are the mass and the charge of an electron respectively, v the average velocity and τ the relaxation time. τ is the time an electron needs to bring its average velocity to zero due to scattering in the crystal lattice of the conductor. At equilibrium ($\partial v/\partial t = 0$), the average velocity v results from the opposite motion of the scattering and the acceleration of the electrons by an electrical field. It is given by

$$\vec{v} = \frac{e\vec{E}\tau}{m} \tag{A.2}$$

Considering n electrons per unit volume, this leads to Ohm‘s law

$$\vec{J} = ne\vec{v} = \frac{ne^2\tau}{m}\vec{E} = \sigma\vec{E} \tag{A.3}$$

For a constant electrical field we get therefore $\sigma = ne^2\tau/m$. Applying an alternating electrical field in the form of $\vec{E} = \vec{E}_0 e^{j\omega t}$, the conductivity gets complex. From the Drude model described above this leads to the complex conductivity for normal conductors

$$\sigma = \frac{ne^2\tau}{m(1+\omega^2\tau^2)} - j\frac{ne^2\omega\tau^2}{m(1+\omega^2\tau^2)} = \sigma_{1n} - j\sigma_{2n} \tag{A.4}$$

For normal conductors even at microwave frequencies, the imaginary part of σ can be neglected due to the fact that $\omega << 1/\tau$.
The theory of superconductivity predicts a zero dc-resistance of the material below its critical temperature T_C. In that case we have no more scattering effects of electrons in the crystal lattice which gave rise to the famous London theory [33]. To apply the model of conductivity of normal conductors to the superconducting case, we have to modify A.1. The relaxation time τ has a finite value for normal conductors due to the interaction of electrons and the crystal lattice as described above. If there is no more scattering in the lattice, we can assume that $\tau = \infty$. This simplifies the equation of motion to

$$m\frac{\partial\vec{v}}{\partial t} = e\vec{E} \tag{A.5}$$

Applying an ac-electrical field analog to the normal conducting case leads to *first London equation*

$$\frac{\partial\vec{J}_s}{\partial t} = \frac{n_s e^2}{m}\vec{E} \tag{A.6}$$

where n_s is the density of superconducting carriers and J_s is the superconducting current density. This equation describes the zero resistance effect and the resulting super-current of superconductors. As for the normal conducting case we can extract the conductivity, which is purely reactive for superconductors.

$$\sigma = -j\frac{n_s e^2}{\omega m} \tag{A.7}$$

A second effect of superconductors that was observed is the complete diamagnetism that is not a direct consequence of the perfect conductivity. It describes a complete expulsion of magnetic flux from the bulk superconductor by generating opposite currents at the surface of the superconductor to cancel out any flux that penetrates into it. This effect, the so-called *Meissner-effect* [40] was first discovered by Meissner and Ochsenfeld in 1933 and was later described by the London bothers with the *second London equation*

$$\nabla \times \vec{J_s} = -\frac{\mu_0 n_s e^2}{m}\vec{H} \tag{A.8}$$

If we define $\lambda_L = \sqrt{m/\mu_0 n_s e^2}$, which is the so called London penetration depth, we can rewrite the two London equations to

$$1.\ London\ equation: \quad \frac{\partial \vec{J_s}}{\partial t} = \frac{1}{\mu_0 \lambda_L^2}\vec{E} \tag{A.9}$$

$$2.\ London\ equation: \ \nabla \times \vec{J_s} = -\frac{1}{\lambda_L^2}\vec{H} \tag{A.10}$$

Applying a magnetic field parallel to the surface of the superconductor and using the second London equation leads to a magnetic field density at the surface given by

$$\vec{H}(x) = \vec{H}(0)e^{-\frac{x}{\lambda_L}} \tag{A.11}$$

where x is the distance from the surface of the superconductor. The London penetration depth λ_L describes the distance in the bulk from the edge of the superconductor, where the magnetic field is reduced by a factor of e. For bulk aluminum for example $\lambda_L \approx 50\ nm$ [50]. This penetration of the magnetic field into the superconductor is demonstrated in Fig. A.1. A more detailed derivation of the two London equations and the penetration depth λ_L and the London equations can be found in [16].
The description of electrodynamics of superconductors by the London theory is a phenomenological approach and is only valid for thick bulk superconductors. For

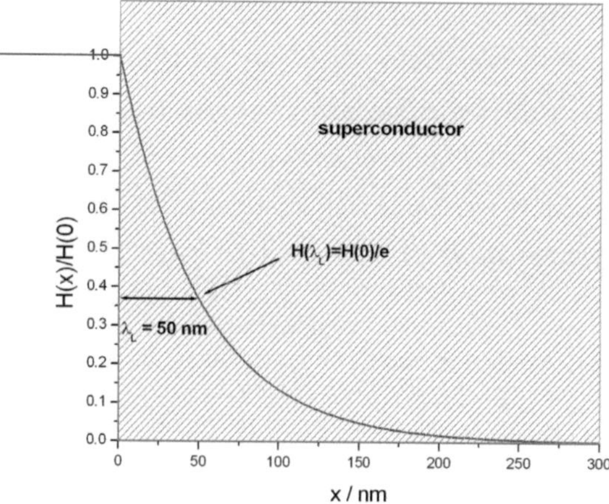

Fig. A.1.: Penetration of magnetic fields in superconductors described by the London penetration depth λ_L. At this distance, the magnetic field is reduced by a factor of e.

very thin films ($d < \lambda_L$) we have to take a more fundamental theory which will be discussed in the following sections.

The two fluid model The concept of a two fluid model to explain the properties of superconductors existed already before the famous BCS theory (see later). Gorter and Casimir [27] developed the most successful theory. Their model describes the coexistence of superconducting carriers, later known as cooper pairs, with density n_s and the thermally excited electrons, so-called quasi-particles with density n_n. A schematic of this two fluid model expressed by different conductivities is shown in Fig. A.2. $\sigma_1(n_n)$ represents the real part of the conductivity due to quasi-particles and $\sigma_2(n_n)$ the imaginary part. $\sigma_2(n_s)$ stands for the conductivity due to cooper pairs and is purely reactive. The conductivities seen in Fig. A.2 depend on the ratio of cooper pairs and quasi-particles, which depends on temperature. Gorter and Casimir empirically found a relation between n_s, n_n and the temperature T.

$$\frac{n_s}{n} = 1 - (\frac{T}{T_C})^4 \qquad \text{(A.12)}$$

with $n = n_s + n_n$ the total charge carrier density. This can directly be applied to the

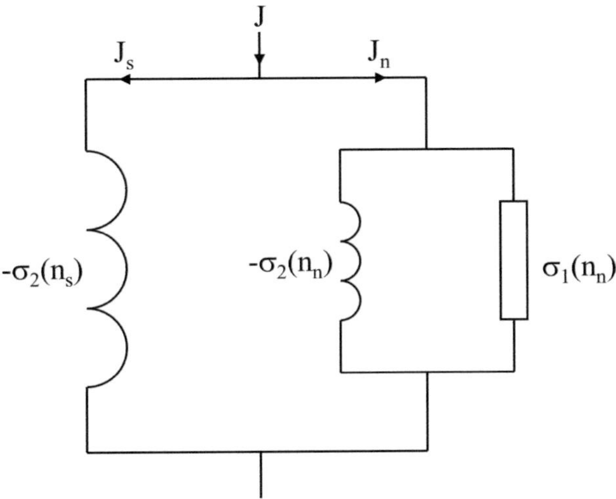

Fig. A.2.: Equivalent circuit describing the two fluid model.

London penetration depth leading with

$$\lambda_L(0) = \sqrt{\frac{m}{n_s e^2 \mu_0}} \tag{A.13}$$

to the temperature dependent London penetration depth

$$\lambda_L(T) = \frac{\lambda_L(0)}{\sqrt{1-(\frac{T}{T_C})^4}} \tag{A.14}$$

For $T \to 0$ λ_L reaches its minimum value $\lambda_L(0)$ due to the domination of cooper pairs, whereas it will increase for increasing temperatures. If the temperature reaches the critical temperature T_C, λ_L becomes infinite because all cooper pairs are broken up into quasi particles. At this point the magnetic field penetrates into the whole thickness of the superconductor and there are no more surface currents excited to expel the magnetic flux. This is described by the second London equation explained above and this behavior has of course a big influence on the conductivity of superconductors.

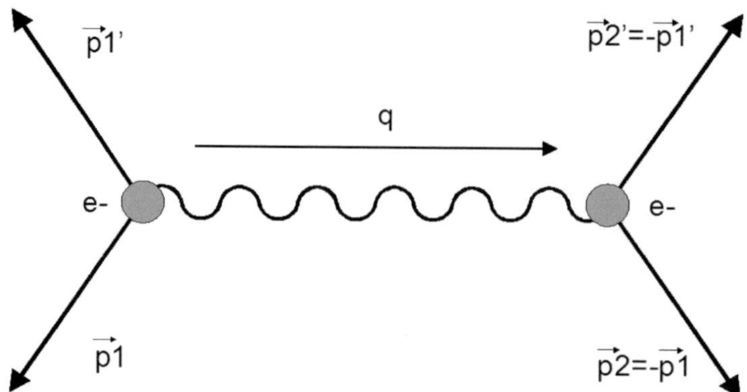

Fig. A.3.: Cooper pair formation due to the interaction of two single electrons with opposite momentum. q describes the positive charge due to a local lattice polarization

The BCS theory The first microscopic theory of superconductivity was presented by **B**ardeen, **C**ooper and **S**chriefer in 1957, which is called the BCS theory. The theory is based on the fact that electrons in a superconductor form cooper pairs at $T < T_C$. If an electron passes the lattice of a metal it can distort the ions in the crystal lattice, which causes a local polarization. Such a system of an electron interacting with lattice ions is called a Polaron. This local lattice polarization acts as a positive charge q, attracting a second electron. This is only possible for electrons with opposite momentum $\vec{p}$ and spin $\vec{s}$. A model of this attraction of two electrons in a crystal lattice is shown in Fig. A.3. The maximum distance over which the binding of the two electrons is effective is called the *coherence length* of a cooper pair. It is a material dependent parameter and can be calculated from the BCS theory to

$$\xi_0 = 0.18 \frac{\hbar v_f}{k_B T_C} \tag{A.15}$$

where v_f is the Fermi velocity. Typical superconductors exhibit a ξ of around 1 μm such as aluminum with $\xi \approx 1700$ nm [50]. Pairs of electrons bounded by this phenomenon are called cooper pairs [14]. To conserve the overall momentum in the system, $\vec{p}1 + \vec{p}2 = \vec{p}1' + \vec{p}2'$ has to be guaranteed. One can describe this dynamic polarization by the formation of virtual phonons. The phonons due to the interaction of two electrons exist only for a short period of time, which gives them the name virtual phonons. Due to the Heisenberg uncertainty principle, this short period of time can cause small changes in Energy ($\Delta E \geq \hbar/\Delta t$). Another difference to classical electrons is the spin of a cooper pair. The spins ($s\uparrow, s\downarrow$) of the two electrons compensate themselves to zero, which means, that a cooper pair cannot be considered as a Fermion but as boson and follows therefore the Bose-Einstein statistics. The Pauli exclusion does not apply to bosons and therefore more than one cooper pair can be in the same quantum state. Approaching $T = 0K$ all cooper pairs can condensate in a single quantum state, where their energy is less than two times the Fermi-energy E_F. This amount of energy is called the gap energy Δ of an superconductor and is formed on either side of the Fermi-energy. Therefore an energy of Δ is needed to break a cooper pair or to excite quasi-particles. This parameter is temperature dependent and reaches its maximum $\Delta(0)$ at $T = 0\ K$, where all electrons form cooper pairs. At $T = 0\ K$ the energy gap can be approximated with

$$2\Delta(0) = 3.526 k_B T_C \tag{A.16}$$

For higher temperatures, close to T_C, more and more cooper pairs break apart leading to an increase in quasi-particles and the energy gap approaches zero. At this transition there are no more cooper pairs left and the material becomes normal conductive. The temperature dependence of Δ is given by [60, 63]

$$\frac{1}{N(0)V} = \int_{-\hbar\omega_D}^{\hbar\omega_D} \frac{tanh \frac{\sqrt{E^2+\Delta(T)^2}}{2k_B T}}{2\sqrt{E^2+\Delta(T)^2}} dE \tag{A.17}$$

where N(0)V represents the total phonon interaction potential and ω_D is the Debye frequency. At temperatures $T << T_C$ the energy gap varies very little and is close to $\Delta(0)$. For $T \approx T_C$ it can be approximated by

$$\Delta(T) = \Delta(0)\sqrt{1 - \frac{T}{T_C}} \tag{A.18}$$

From this we can calculate the superconducting density of states as described in [60]

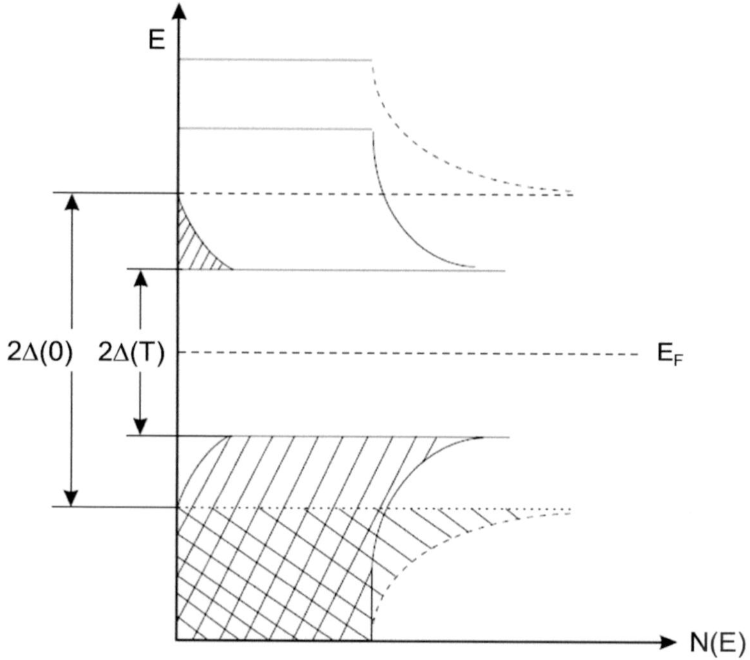

Fig. A.4.: Density of states of cooper pairs and quasi-particles depending on the temperature T.

to

$$N_s = \frac{N(0)E}{\sqrt{E^2 + \Delta(T)^2}} \tag{A.19}$$

$N(0)$ is the single spin density of states.

In [60] the model of quasi-holes and quasi-particles from the semiconductor model is used to describe the number of excited electrons to quasi-particles. The principle of the quasi-particle excitation depending on the temperature and therefore on the gap energy is shown in Fig. A.4. For $T = 0\ K$ the energy gap reaches its maximum and all electrons form cooper pairs with a state energy below $E = E_F - \Delta(0)$. If the temperature increases, Δ becomes smaller and quasi-particles are excited leaving quasi-holes in the band below $E = E_F - \Delta(0)$. The quasi-particles add now a normal state conductivity σ_1 to the total conductivity. The density of excited quasi-particles at a temperature T can be calculated with

$$n_{qp}(T) = 4\int_{\infty}^{0} f(E)\frac{tanh\frac{\sqrt{E^2+\Delta(T)^2}}{2k_BT}}{2\sqrt{E^2+\Delta(T)^2}}dE \tag{A.20}$$

where $f(E) = 1/(1+e^{E/k_BT})$ is the Fermi-distribution function.

A.2. Analytical calculation of L and C as a first estimation

Interdigital capacitor The calculation of the capacity for a microstrip interdigital capacitor, as shown in Fig. A.5 is described in [65]. The serial capacitance can hereby be approximated with

$$C = \frac{\varepsilon_r + 1}{B}((n-3)A_1 + A_2)L_c^2 \quad [pF] \tag{A.21}$$

with

$$A_1 = \left[0.3349057 - 0.15287116\left(\frac{d}{\mathrm{w}}\right)\right]^2 \tag{A.22}$$

and

$$A_2 = \left[0.50133101 - 0.22820444\left(\frac{d}{\mathrm{w}}\right)\right]^2 \tag{A.23}$$

where ε_r is the dielectric constant of the substrate, B the total width of the capacitor structure, n the number of fingers, w the width of a finger, d the conductor thickness and L_c the finger length. This equation is valid for $L_c > B/n$ and gives a good estimation for the capacity of the LEKID interdigital capacitor.

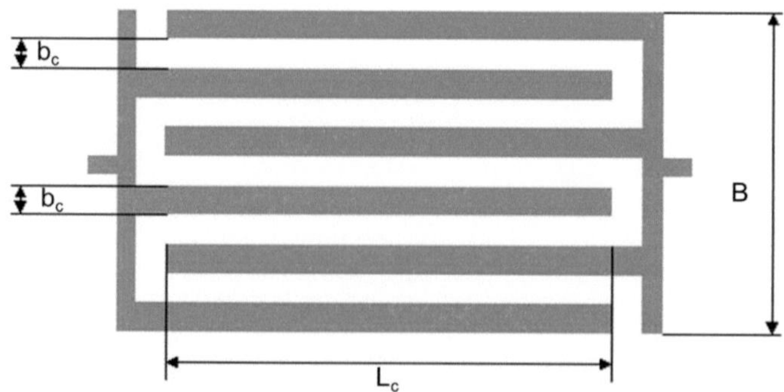

Fig. A.5.: Schematic of a six finger microstrip interdigital capacitor.

Inductive meander line The meander geometry of the NIKA 2.05 mm LEKID consists of 8 parallel wires of length l_1 and 7 short interconnections of length b_i. We can consider the 8 parallel lines, as a first approach, as 2 parallel wires each with length $l_2 = (8 \cdot l_1)/2$ and opposite current directions giving the inductance L_1. The inductance in the short interconnections L_2 is simply the sum of the inductances of each segment with total length $l_3 = 7 \cdot b_i$. The inductance L_{line} is then given by

$$L_{line} = L_1 + L_2 + L_{kin,tot} \tag{A.24}$$

where $L_{kin,tot}$ is the kinetic inductance of the total meander length. These three inductances forming the total inductance are shown as a schematic in Fig. A.6 and have to be calculated to determine L_{line}.

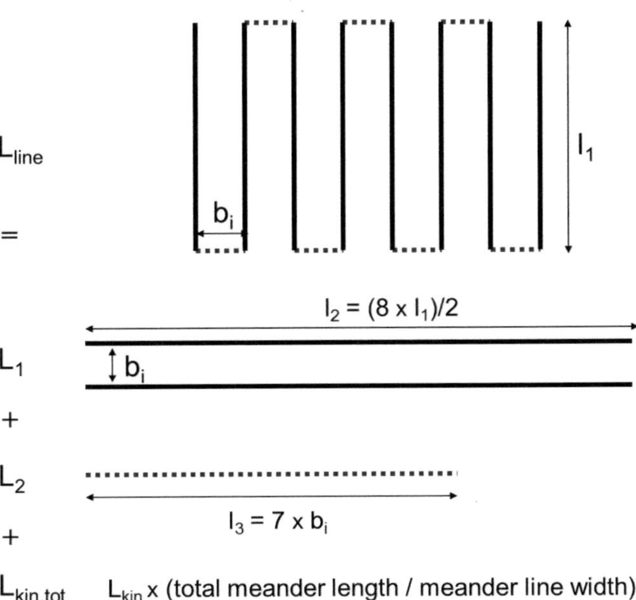

Fig. A.6.: Schematic of the total inductance modeling of the LEKID meander line. The parallel lines (solid) are modeled as two long parallel lines with opposite current directions. The short interconnections (dashed lines) can be considered as one thin wire with total length l_3. The third part is the kinetic inductance that follows from the superconducting properties of the conductor.

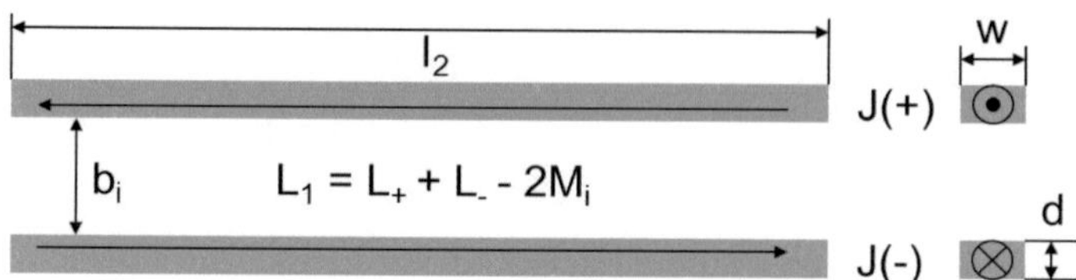

Fig. A.7.: Schematic of two parallel wires with opposite current directions to calculate the inductance of the LEKID meander line, where w is the width and d the thickness of the wire.

The inductance of two parallel wires with opposite current directions (see Fig. A.7) is given by

$$L_1 = L_+ + L_- - 2M_i \tag{A.25}$$

where L_+ and L_- are the self inductances of each wire (+/- are indicating the current direction) and M_i is the mutual inductance that lowers the total inductance L_1 depending on the distance between the wires. L_1 is hereby given by [65, 28]

$$L_1 \approx 4l_2 \left(ln \frac{b_i}{\mathrm{w}+d} + 1.5 \right) \quad [nH] \tag{A.26}$$

where l_2, d, w and b_i are in cm. This equation is valid for $d << \mathrm{w}$.
The inductance of a simple rectangular wire to calculate the sum of the short lines is given by

$$L_2 = 2l_3 \left(ln \frac{2l_3}{\mathrm{w}+d} + 0.5 + 0.2235 \frac{\mathrm{w}d}{l_3} \right) \quad [nH] \tag{A.27}$$

where l_3, d and w are in cm as well.
The kinetic inductance for the entire meander line length is given by

$$L_{kin,tot} = \frac{total\ meander\ length}{meander\ line\ width} L_{kin} \tag{A.28}$$

where L_{kin} is the kinetic inductance per square, whose calculation is described in Chapter 3.

A.3. List of symbols and constants

Acronyms

Name	description
ADC	Analog to digital converter
CMB	Cosmic microwave background
CPW	Coplanar waveguide
DAC	Digital to analog converter
FDM	Frequency division multiplexing
FPGA	Field programmable gate array
FTS	Fourier transform spectroscopy
FWHM	Full wave half maximum
IQ-mixer	Inphase-quadrature mixer
IRAM	Institut de Radioastronomie millimétrique
KID	Kinetic inductance detector
LEKID	Lumped element kinetic inductance detector
NEFD	Noise equivalent flux density
NEP	Noise equivalent power
NET	Noise equivalent temperature
NIKA	Néel IRAM KID array
PSF	Point spread function
RIE	Reactive ion etching
SNR	Signal to noise ratio
S-parameter	Scattering parameter
TES	Transition edge sensor

Symbols

Name	Variable	unit
Kinetic energy	E_{kin}	J
Magnetic energy	E_m	J
Focal length	F	m
resonance frequency	f_{res}	Hz
Geometrical inductance	L_{geo}	H
Kinetic inductance	L_{kin}	H
Magnetic inductance	L_m	H
Mutual inductance	M	H
charge carrier density	n	m^{-3} m^{-3}
Single spin density	N(0)	$m^{-3}eV^{-1}$
Normal conducting charge carrier density	n_n	m^{-3}
Number of quasiparticles	N_{qp}	1
Superconducting charge carrier density	n_s	m^{-3}
Intrinsic or unloaded quality factor	Q_0	1
Coupling quality factor	Q_C	1
Loaded quality factor	Q_L	1
Sheet resistance	$R_\square$	Ω/square
Spectral noise density	S_n(f)	Hz/$\sqrt{Hz}$
Critical temperature	T_C	K
average charge carrier velocity	v_0	m/s
minimum waist	$w_{o1,o2}$	m
Free space impedance	Z_0	Ω
Confocal distance	z_{c1}	m
Characteristic line impedance	Z_L	Ω
Surface impedance	Z_s	Ω/square
detector frequency response	ΔS	Hz

Greek letters

Name	Variable	unit
Kinetic inductance fraction	α	1
Gap energy of superconductor	Δ	J
Loss tangent	$\tan\delta$	1
Relative permittivity	ε_r	1
Effective relative permittivity	$\varepsilon_{r,eff}$	1
Coupling factor	κ	1
Wavelength	λ	m
Wavelength in vacuum	λ_0	m
London penetration depth	λ_L	m
Specific resistivity	ρ	Ωm
Complex conductivity	$\sigma = \sigma_1 - j\sigma_2$	S/m
Electron-phonon relaxation time	τ	s
Quasiparticle lifetime	τ_{qp}	s
Angular frequency	ω=2πf	Hz
Solid angle	Ω	sr
Apex angle	ψ	deg
Emission angle	β	deg
Coherence length	ξ_0	m

Fundamental constants [42, 43]

Name	Constant	Value	unit
Speed of light in vacuum	c	2.997956×10^{8}	m/s
Elementary charge	e	$1.602176565 \times 10^{-19}$	C
Planck constant	h	$6.62606957 \times 10^{-34}$	J·s
Boltzmann constant	k_B	$1.3806488 \times 10^{-23}$	J/K
Electron mass	m_e	$9{,}109382 \times 10^{-31}$	kg
Vacuum permittivity	ε_0	$8.854187817 \times 10^{-12}$	F/m
Vacuum permeability	μ_0	$4 \cdot \pi \cdot 10^{7} = 1.256637061 \times 10^{-6}$	H/m
Pi	π	3.141592653	1
Free space impedance	Z_0	376.730313461	Ω

Literaturverzeichnis

[1] Z. 12. Optical and Illumination Design Software. *Redmond, WA 98053, USA (www.zemax.com)*, 2012.

[2] M. Afsar, M. Obol, K. Korolev, and S. Naber. The use of microwave, millimeter wave and terahertz spectroscopy for the detection, diagnostic and prognosis of breast cancer. *33rd International Conference on Infrared, millimeter and Terahertz waves, IRMMW-THz*, pages 1–3, 2008.

[3] W. Altenhoff, C. Thum, and H. Wendker. Radio-emission from stars - a survey at 250 GHz. *Astronomy and Astrophysics*, 281(1):161–183, 1994.

[4] R. Arendt, J. George, J. Staguhn, D. Benford, M. Devlin, S. Dicker, D. J. Fixsen, K. Irwin, C. Jhabvala, P. Korngut, A. Kovács, S. Maher, B. Mason, T. Miller, S. Moseley, S. Navarro, A. Sievers, J. Sievers, E. Sharp, and E. Wollack. The Radio-2 mm spectral index if the Crab Nebula measured with GISMO. *Astrophysical Journal*, 734(1), 2011.

[5] A. Baryshev, J. J. A. Baselmans, A. Freni, G. Gerini, H. Hoevers, A. Iacono, and A. Neto. Progress in Antenna Coupled Kinetic Inductance Detectors. *IEEE Transactions on Terahertz Science and Technology*, 1(1):112–123, 2011.

[6] C. Bennett, M. Bay, M. Halpern, G. Hinshaw, C. Jackson, N. Jarosik, A. Kogut, M. Limon, S. Meyer, L. Page, D. Spergel, G. Tucker, D. Wilkinson, E. Wollack, and E. Wright. The Microwave Anisotropy Probe (MAP) Mission. *The Astrophysical Journal*, 583:1, 2003.

[7] C. Bertout, F. Combes, A. Ferrara, T. Forveille, S. N. Shore, E. Tolstoy, and C. Walmsley. Early results from the Planck mission. *Astronomy and Astrophysics*, 536(E1), 2011.

[8] A. Bideaud. Développement d'une caméra pour la radioastronomie millimétrique. *PhD thesis, Université Joseph Fourier, Grenoble, France (in French)*, 2010.

[9] B. Boggess, J. Mather, R. Weiss, C. Bennett, E. Cheng, E. Dwek, S. Gulkis, M. Hauser, M. Janssen, T. Kelsall, S. Meyer, S. Moseley, T. Murdock, R. Shafer, R. Silverberg, G. Smoot, D. Wilkinson, and E. Wright. The COBE mission: its design and performance two years after launch. *The Astrophysical Journal*, 397:420–429, 1992.

[10] O. Bourrion, A. Bideaud, A. Benoit, A. Cruciani, J. Macias-Perez, A. Monfardini, M. Roesch, L. Swenson, and C. Vescovi. Electronics and data acquisition demonstrator for a kinetic inductance camera. *Journal of instrumentation*, 6(P06012), 2011.

[11] J. Bray and L. Roy. Measuring the unloaded, loaded, and external quality factors of one- and two-port resonators using scattering-parameter magnitudes at fractional power levels. *IEE Proceedings - Microwaves, Antennas and Propagation*, 151(4):345–350, 2004.

[12] J. E. Carlstrom, G. P. Holder, and E. D. Reese. Cosmology with the Sunyaev-Zel'dovich effect. *Annual Review of Astronomy and Astrophysics*, 40:643–680, 2002.

[13] R. Chambers. The anomalous skin effect. *Proceedings of the Royal Society of London, Series A, Mathematical and Physical Sciences*, 215(1123):481–497, 1952.

[14] L. N. Cooper. Bound electron pairs in a degenerate Fermi gas. *Physical Review*, 104(4):1189–1190, 1956.

[15] P. Day, H. LeDuc, B. Mazin, A. Vayonakis, and J. Zmuidzinas. A broadband superconducting detector suitable for use in large arrays. *Nature*, 425:817–821, 2003.

[16] S. Doyle. Lumped Element Kinetic Inductance Detectors. *PhD thesis, Cardiff University, Cardiff, UK*, 2008.

[17] S. Doyle, P. Mauskopf, J. Naylon, A. Porch, and C. Duncombe. Lumped element kinetic inductance detectors. *Journal of Low Temperature Physics*, 151(1-2):2–5, 2008.

[18] T. Durand. Réalisation d'un interféromètre de Martin Puplett pour le dévelpement d'une caméra bolométrique. *PhD thesis, Université Joseph Fourier, Grenoble, France (in French)*, 2007.

[19] P. Encrenaz, C. Laurent, S. Gulkis, E. Kollberg, and W. Winnesser. Coherent detection at millimeter wavelengths and their applications. *Nova Science Publisher, New York*, 1991.

[20] A. B. et al. Archeops: a high resolution, large sky coverage balloon experiment for mapping cosmic microwave background anisotropies. *Physical Review*, 17(2):101–124, 2002.

[21] S. M. et al. Instrument, Method, Brightness and Polarization Maps from the 2003 flight of BOOMERANG. *Astronomy and Astrophysics*, 458:687–716, 2003.

[22] J. Gao. The physics of superconducting microwave resonators. *PhD thesis, Caltech, Pasadena, USA*, 2008.

[23] J. Gao, J. Zmuidzinas, B. Mazin, H. LeDuc, and P. Day. Noise properties of superconducting coplanar waveguide microwave resonators. *Applied Physics letters*, 90(102507), 2007.

[24] B. Geilikman and V. Kresin. Kinetic and non-steady effects in superconductors. *John Wiley and sons*, 1974.

[25] J. Glenn, J. Bock, G. Chattopaadhyay, S. Edgington, A. Lange, J. Zmuidzinas, P. Mauskopf, B. Rownd, L. Yuen, and P. Ade. Bolocam: a millimeter-wave bolometric camera. *Proceedings SPIE*, 3357(326), 1998.

[26] P. Goldsmith. Quasioptical systems. *IEEE Press/Chapman & Hall Publishers Series on Microwave Technology and RF*, 1997.

[27] C. J. Gorter and H. B. G. Casimir. The thermodynamics of the superconducting state. *Physic. Z.*, 15:539–542, 1956.

[28] F. Grover. Inductance calculations. *Van Nostrand, Princton, NJ, 1946; reprinted by Dover Publications, NY*, 1962.

[29] K. Irwin and G. Hilton. Cryogenic particle detection. *Topics in Applied Physics, C. Enss, ed., Springer, Berlin, Germany*, 99:63–149, 2005.

[30] S. Kaplan, C. Chi, D. Langenberg, J. Chang, S. Jafarey, and D. Scalapino. Quasiparticle and phonon lifetimes in superconductors. *Phys. Rev. B*, 14::4854–4873, 1976.

[31] A. Khanna and Y. Gerault. Determination of loaded, unloaded and external quality factors of a dielectric resonator coupled to a microstrip line. *IEEE Transactions on Microwave Theory and Techniques*, 31(3):261–264, 1983.

[32] S. Lee, J. Gildemeister, W. Holmes, A. Lee, and P. Richards. Voltage biased superconducting transition edge bolometer with strong electrothermal feedback operated at 370mK. *Applied Optics*, 37(16):3391–3397, 1998.

[33] F. London and H. London. The Electromagnetic Equations of the Superconductor. *Proceedings of the Royal Society Series*, A(149):71–88, 1935.

[34] P. Maloney, N. Czakon, P. Day, T. Downes, R. Duan, J. Gao, J. Glenn, S. Golwala, M. Hollister, H. LeDuc, B. Mazin, O. Noroozian, H. Nguyen, J. Sayers, J. Schlaerth, S. Siegel, J. Vaillancourt, A. Vayonakis, R. Wilson, and J. Zmuidzinas. MUSIC for sub/millimeter astrophysics. *Proceedings SPIE*, 7741(77410F), 2010.

[35] B. Mandelbrot. The fractal geometry of nature. *Freeman, New york*, 1983.

[36] F. Mattiocco and M. Carter. 80 - 360 GHz very wide band millimeter wave network analyzer. *International journal of infrared millimeter waves*, vol. 16(no. 12), 1995.

[37] D. C. Mattis and J. Bardeen. Theory of the anomalous skin effect in normal and superconducting metals. *Physical Review*, 111:412–417, 1956.

[38] T. May, G. Ziegler, S. Anders, V. Zakosarenko, M. Starkloff, H.-G. Meyer, G. Thorwirth, and E. Kreysa. Passive stand-off terahertz imaging with 1 hz frame rate. *Proceedings of SPIE*, 6949(69490C), 2008.

[39] B. Mazin. Microwave kinetic inductance detectors. *dissertation (PhD), Caltech,Pasadena, USA*, 2005.

[40] W. Meissner and R. Ochsenfeld. Ein neuer Effekt bei Eintritt der Supraleitfï¿$\frac{1}{2}$higkeit. *Naturwissenschaften*, 21(44):787–788, 1933.

[41] C. Microwave Studio. Computer Simulation Technology. *Wellesley Hills, MA, USA (www.cst.com)*, 2011.

[42] P. Mohr and B. Taylor. CODATA recommended values of fundamental physical constants. *Reviews of Modern Physics*, 80(2):633–730, 2008.

[43] P. Mohr and B. Taylor. The NIST reference on constants, units and uncertainty: CODATA recommended values of fundamental physical constants. *Website:http://physics.nist.gov/constants*, 2012.

[44] A. Monfardini, A. Benoit, A. B. L. Swenson, A. Cruciani, P. Camus, C. Hoffmann, F. Desert, S. Doyle, P. Ade, P. Mauskopf, C. Tucker, M. Roesch, S. Leclercq, K. Schuster, A. Endo, A. Baryshev, J. Baselmans, L. Ferrari, S. Yates, O. Bourrion, J. Macias-Perez, C. Vescovi, M. Calvo, and C. Giordano. A dual band millimeter-wave kinetic inductance camera for the IRAM 30m telescope. *Astrophysical Journal Supplements*, 194(2), 2011.

[45] A. Monfardini, L. Swenson, A. Bideaud, F. Desert, S. Yates, A. Benoit, A. Baryshev, J. Baselmans, S. Doyle, B. Klein, M. Roesch, C. Tucker, P. Ade, M. Calvo, P. Camus, C. Giordano, R. Guesten, C. Hoffmann, S. Leclercq, P. Mauskopf, and K. Schuster. NIKA: a millimeter-wave kinetic inductance camera. *Astronomy and Astrophysics*, 521:A29, 2010.

[46] O. Noroozian, J. Gao, J. Zmuidzinas, H. Leduc, and B. Mazin. Two-level system noise reduction for Microwave Kinetic Inductance Detectors. *AIP Conference Proceedings LTD13*, 1185:148–152, 2009.

[47] A. Omont, P. Cox, F. Bertoldi, R. McMahon, C. Carilli, and K. Isaak. A 1.2 mm MAMBO/IRAM-30 m survey of dust emission from the highest redshift PSS quasars. *Astronomy and Astrophysics*, 374(2), 2001.

[48] A. B. Pippard. An experimental and theoretical study of the relation between magnetic field and current in a superconductor. *Proc. Roy. Soc.*, A216:547–568, 1953.

[49] D. M. Pozar. Microwave engineering. *John Wiley and Sons,Inc.*, 1998.

[50] R. Pï¿½pel. Surface impedance and reflectivity of superconductors. *Journal of Applied Physics*, 66(12):5950–5957, 1989.

[51] P. Richards. Bolometers for infrared and millimeter waves. *Journal of Applied Physics*, 76(1), 1994.

[52] H. Sagan. Space-filling curves. *New York: Springer Verlag*, 1994.

[53] J. Schlaerth, N. Czakon, P. Day, T. Downes, R. Duan, J. Gao, J. Glenn, S. Golwala, M. Hollister, H. LeDuc, B. Mazin, P. Maloney, O. Noroozian, H. Nguyen, J. Sayers, S. Siegel, J. Vaillancourt, A. Vayonakis, R. Wilson, and J. Zmuidzinas. MKID multicolor array status and results from DemoCam. *Proceedings SPIE*, 7741(774109), 2010.

[54] J. Schlaerth, A. Vayonakis, P. Day, J. Glenn, J. Gao, S. Golwala, S. Kumar, H. LeDuc, J. Vaillancourt, and J. Zmuidzinas. A millimeter and submillimeter kinetic inductance detector camera. *Journal of low temperature physics*, 151(3-4):684–689, 2008.

[55] K.-F. Schuster, C. Boucher, W. Brunswig, M. Carter, J.-Y. Chenu, B. Foullieux, A. Greve, D. John, B. Lazareff, S. Navarro, A. Perrigouard, J.-L. Pollet, A. Sievers, C. Thum, and H. Wiesemeyer. A 230 ghz heterodyne receiver array for the IRAM 30 m telescope. *Astronomy and Astrophysics*, 423:1171–1177, 2004.

[56] A. Sergeev, V. Mitin, and B. Karasik. Ultrasensitive hot-electron kinetic-inductance detectors operating well below the superconducting transition. *Applied Physics letters*, 80((5)):817–819, 2002.

[57] S. Software. em user's manual and xgeom user's manual.

[58] J. Staguhn, Benford, R. Arendt, D. Fixsen, A. Karim, A. Kovacs, S. Leclercq, S. Mahera, T. Miller, S. Moseley, and E. Wollack. Latest results from GISMO: a 2-mm bolometer camera for the IRAM 30-m telescope. *EAS Publications Series*, 52:267–271, 2012.

[59] L. Swenson, A. Cruciani, A. Benoit, M. Roesch, C. Yung, A. Bideaud, and A. Monfardini. High-speed phonon imaging using frequency-multiplexed kinetic inductance detectors. *Applied Physics letters*, 96(26), 2010.

[60] M. Tinkham. Introduction to superconductivity, 2nd edition. *McGraw-Hill International Editions, Singapore*, 1996.

[61] R. Ulrich. Far-infrared properties of metallic mesh and its complementary structure. *Infrared Physics*, vol. 7:pp. 37–55, 1967.

[62] R. Ulrich, T. Bridges, and M. Pollack. Variable metal mesh coupler for far infrared lasers. *Applied Optics*, vol. 9(no. 11):pp. 2511–2516, 1970.

[63] T. van Tuzer and C. W. Turner. Principles of superconductive devices and circuits. *Elsevier North Holland, Inc, NY*, 1981.

[64] K. Vinoy, K. Jose, V. Varadan, and V. Varadan. Hilbert curve fractal antenna: a small resonant antenna for VHF/UHF applications. *Microwave and optical technology*, 29(4), 2001.

[65] B. Wadell. Transmission line design handbook. *Teradym, Inc, Boston, MA*, 1991.

[66] L. Whitbourn and R. Compton. Equivalent-circuit formulas for metal grid reflectors at a dielectric boundary. *Applied Optics*, vol. 24(no. 2):pp. 217–220, 1985.

[67] A. Woodcraft, P. Ade, D. Bintley, J. House, C. Hunt, R. Sudiwala, W. Doriese, W. Duncan, G. Hilton, K. Irwin, C. Reintsema, J. Ullom, M. Audley, M. Ellis, W. Holland, M. MacIntosh, C. Dunare, W. Parkes, A. Walton, J. Kycia, M. Halpern, and E. Schulte. Electrical and optical measurements on the first SCUBA-2 prototype 1280 pixel submillimeter superconducting bolometer array. *Rev. Sci. Instrum.*, 78(2):024502, 2007.

[68] Z. Wu and L. Davis. Surface roughness effect on surface impedance of superconductors. *Journal of Applied Physics*, 76(6):3669–3672, 1994.

[69] J. Yates, A. Baryshev, J. Baselmans, B. Klein, and R. Gï¿½sten. Fast Fourier transform spectrometer readout for large arrays of microwave kinetic inductance detectors. *Applied Physics letters*, 95(4), 2009.

[70] J. Zhu, A. Hoorfar, and N. Engheta. Bandwidth, cross-polarization and feed-point characteristics of matched Hilbert antennas. *IEEE Antennas and wireless propagation letters*, 2:2–5, 2003.

[71] J. Zhu, A. Hoorfar, and N. Engheta. Peano antennas. *IEEE Antennas and wireless propagation letters*, 3:71–74, 2004.

Karlsruher Schriftenreihe zur Supraleitung (ISSN 1869-1765)

Herausgeber: Prof. Dr.-Ing. M. Noe, Prof. Dr. rer. nat. M. Siegel

Die Bände sind unter www.ksp.kit.edu als PDF frei verfügbar oder als Druckausgabe bestellbar.

Band 001
Christian Schacherer
Theoretische und experimentelle Untersuchungen zur Entwicklung supraleitender resistiver Strombegrenzer. 2009
ISBN 978-3-86644-412-6

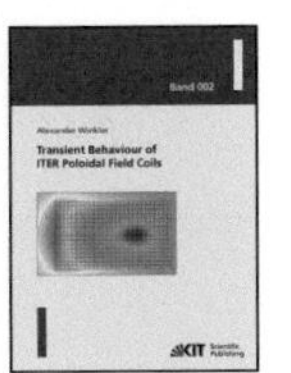

Band 002
Alexander Winkler
Transient behaviour of ITER poloidal field coils. 2011
ISBN 978-3-86644-595-6

Band 003
André Berger
Entwicklung supraleitender, strombegrenzender Transformatoren. 2011
ISBN 978-3-86644-637-3

Band 004
Christoph Kaiser
High quality Nb/Al-AlOx/Nb Josephson junctions. Technological development and macroscopic quantum experiments. 2011
ISBN 978-3-86644-651-9

Band 005
Gerd Hammer
Untersuchung der Eigenschaften von planaren Mikrowellen-resonatoren für Kinetic-Inductance Detektoren bei 4,2 K. 2011
ISBN 978-3-86644-715-8

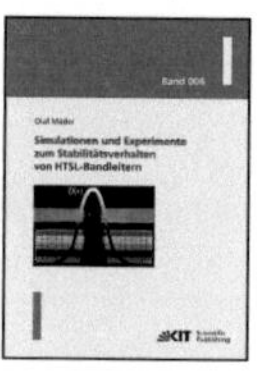

Band 006
Olaf Mäder
Simulationen und Experimente zum Stabilitätsverhalten von HTSL-Bandleitern. 2012
ISBN 978-3-86644-868-1

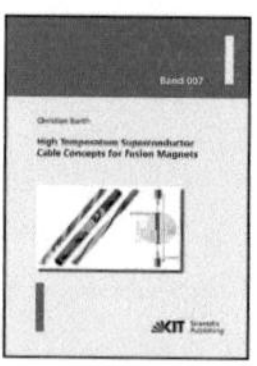

Band 007
Christian Barth
High Temperature Superconductor Cable Concepts for Fusion Magnets. 2013
ISBN 978-3-7315-0065-0

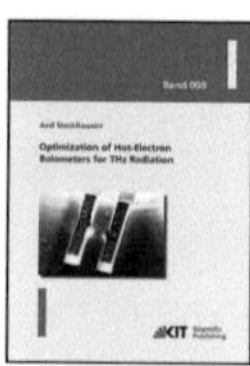

Band 008
Axel Stockhausen
Optimization of Hot-Electron Bolometers for THz Radiation. 2013
ISBN 978-3-7315-0066-7

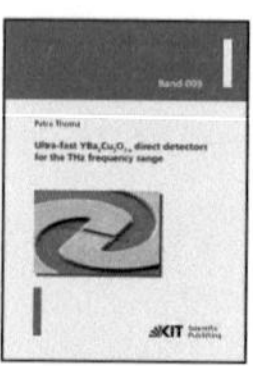

Band 009
Petra Thoma
Ultra-fast $YBa_2Cu_3O_{7-x}$ direct detectors for the THz frequency range. 2013
ISBN 978-3-7315-0070-4

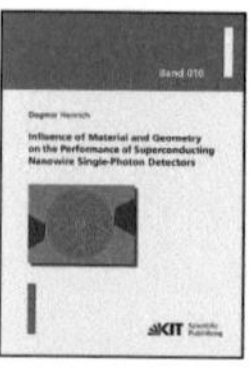

Band 010
Dagmar Henrich
Influence of Material and Geometry on the Performance of Superconducting Nanowire Single-Photon Detectors. 2013
ISBN 978-3-7315-0092-6

Band 011
Alexander Scheuring
Ultrabreitbandige Strahlungseinkopplung in THz-Detektoren. 2013
ISBN 978-3-7315-0102-2

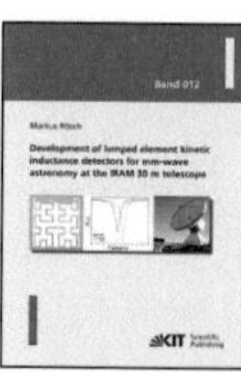

Band 012
Markus Rösch
Development of lumped element kinetic inductance detectors for mm-wave astronomy at the IRAM 30 m telescope. 2013
ISBN 978-3-7315-0110-7